ON BECOMING
AN ENGINEER

ON BECOMING AN ENGINEER

A Guide to Career Paths

J. David Irwin
Earle C. Williams Eminent Scholar and Head
Electrical Engineering Department
Auburn University

IEEE
PRESS

IEEE Education Society, *Sponsor*

The Institute of Electrical and Electronics Engineers, Inc., New York

This book may be purchased at a discount from the
publisher when ordered in bulk quantities. Contact:

IEEE PRESS Marketing
Attn: Special Sales
445 Hoes Lane, P.O. Box 1331
Piscataway, NJ 08855-1331
Fax: (908) 981-9334

For more information about IEEE PRESS products,
visit the IEEE Home Page: http://www.ieee.org/

Printed in the United States of America
10   9   8   7   6   5   4   3   2   1

ISBN 0-7803-1195-7
IEEE Order Number: PP5380

Library of Congress Cataloging-in-Publication Data

Irwin, J. David (date)
      On becoming an engineer: a guide to career paths / J. David Irwin;
   IEEE Education Society, sponsor.
        p. cm.
      Includes bibliographical references and index.
      ISBN 0-7803-1195-7
      1. Engineering—Vocational guidance. I. IEEE Education Society.
   II. Title.
   TA157.I753 1997
   620′.0023—dc20
                                                                96-32601
                                                                    CIP

Dedicated to my parents
Arthur F. and Virginia F. Irwin

Faraday Lecture Series

The Institute of Electrical and Electronics Engineers, Inc. (IEEE), in conjunction with the Institution of Electrical Engineers (IEE), broadcasts the Faraday Lecture Series in early February of each year. Past Faraday Lectures on telecommunications, cellular communications, and the Eurotunnel were captivating presentations on technological achievement in the field of electrical engineering.

In 1831, Michael Faraday observed the principles of electromagnetic induction. Unlike the earlier discovery that electricity has a magnetic effect, this new observation was not accidental but rather was the result of several years of dedicated and meticulous experimentation. Individual lectures in the series focus on specific aspects of Faraday's important discovery.

The Faraday Lectures, begun in 1924, are designed to introduce young people to the field of electrical engineering through a lively hour-long program which combines background, demonstrations, experiments, and explanations of cutting-edge technology. Schools may receive the hour-long free broadcast via a satellite downlink. Teachers are given a workbook for duplication and distribution to students. Teachers may tape and retain this program indefinitely. Tapes of past broadcasts are also available for purchase.

The Faraday broadcast is intended to build a relationship between the IEEE and the pre-college educational community, and teachers of science, technology, and math. This broadcast highlights the benefits of engineering as a career and the importance of society membership to the engineer. IEEE Sections frequently volunteer to answer students' questions following the broadcast, thus introducing prospective engineering students to potential mentors.

For information on how to receive the Faraday broadcast, contact:

IEEE Educational Activities
445 Hoes Lane
Piscataway, NJ 08855
(908) 562-5485
education.services@ieee.org

BE STRONG AND COURAGEOUS;
DO NOT BE FRIGHTENED OR DISMAYED,
FOR THE LORD YOUR GOD IS WITH YOU
WHEREVER YOU GO.

Joshua 1:9

CHAPTER OPENING PHOTO CREDITS

Chapter 1 Opener: Courtesy of Siemens Corporation

Chapter 2 Opener: Courtesy of United Technologies Pratt & Whitney

Chapter 3 Opener: A bleach kraft mill used in making pulp for paper production. (Courtesy of Kvaerner Pulping Inc.)

Chapter 4 Opener: Courtesy of Siemens Corporation

Chapter 5 Opener: A digester unit for converting wood chips into pulp for making paper. (Courtesy of Kvaerner Pulping Inc.)

Chapter 6 Opener: Courtesy of Siemens Corporation

Chapter 7 Opener: A unit used for recovering chemicals and energy in a pulp mill. (Courtesy of Kvaerner Pulping Inc.)

Chapter 8 Opener: Sunshine Skyway Bridge in Tampa Bay, Florida. (Courtesy of Figg Engineering Group)

Chapter 9 Opener: Courtesy of McDonnell Douglas

Chapter 10 Opener: Courtesy of NASA

Contents

Preface

As we approach the new millennium, we find that one of the most dominant forces in our lives is the rapid and relentless advancement of technology. New products and systems, which result from technology's prodigious march, seem to emerge at a pace that is often difficult for us to assimilate. In addition, many of the problems that face our modern society are extremely complex, and their timely and cost-effective solutions will require the creative energy of a diverse group of dedicated professionals. It is within this exciting environment that engineers play a pivotal role.

The progress of technology, the tentacles of which touch essentially every aspect of our daily lives, has created a sustained demand for engineers. The men and women who enter this profession have before them an exciting career replete with difficult challenges and outstanding opportunities.

This book is designed to describe in some detail the many facets and ramifications involved in the pursuit and attainment of an undergraduate engineering degree. The book should be of interest to prospective engineering students, their parents, guidance counselors, and anyone involved in advising and preparing students for a career in one of the engineering fields. In addition to describing the wide spectrum of engineering careers and the type of work performed by some of the more prominent disciplines, the book presents this proper preparation for an educational program in engineering, a variety of items that enhance the program and one's personal performance in it, as well as a number of critical issues that may play a vital role throughout an engineer's career. Each item is carefully outlined and explained, and references are given for essential information to assist the student and new engineer. Finally, some actual career paths are presented. This latter information was obtained from recent graduates in a variety of engineering disciplines and illustrates the type of work performed by young engineers in their first assignment as engineering professionals.

An attempt has been made to address every aspect of this endeavor, so that those interested in an engineering career will gain an understanding of all that is involved. This knowledge of the total process will hopefully help ensure the successful completion of an education program leading to a degree in engineering.

J. David Irwin
Auburn University

Acknowledgments

I am especially indebted to a number of colleagues for their careful review and very helpful suggestions in the development of this book. They are Dr. Harold N. Conrad, Jr., Assistant to the Dean of Engineering at Auburn University for Student Services; Mr. Gerard H. Gaynor, President, G. H. Gaynor and Associates; Dr. Lawrence P. Grayson, Director, Division of Higher Education Incentive Programs, U.S. Department of Education; Dr. Jerrier A. Haddad, Vice President of IBM, retired; Mr. Edmund K. Miller, Los Alamos National Laboratory, Group Leader, Systems and Robotics, retired; Mr. Richard S. Nichols, Principal Electrical Engineer, CH2M-Hill, retired; Dr. Victor Schutz, Professor of Electrical Engineering, Temple University; and Dr. William F. Walker, Dean of Engineering, Auburn University.

The advice and support of the following individuals is also gratefully acknowledged in the preparation of this manuscript: Dr. Jane Z. Daniels, Director, Women in Engineering Programs, Purdue University; Dr. A. Bruno Frazier, Assistant Professor of Electrical and Biomedical Engineering, University of Utah; and Dr. Edward A. Parrish, President, Worcester Polytechnic Institute.

I am most appreciative of the time and effort that was provided to me by many of my Auburn colleagues: Drs. M. Dayne Aldridge, Larry D. Benefield, Robert P. Chambers, John E. Cochran, Jr., Malcolm A. Cutchins, David F. Dyer, John S. Goodling, Leonard L. Grigsby, Joseph F. Judkins,

Loren D. Lutes, Joe M. Morgan, Stephen B. Seidman, and Vernon E. Unger; Mrs. Dara P. Kloss, Mrs. Raya Zalik, and Ms. Betty Kelley; Messrs. C. D. Ellis and Charles M. Griffin.

Finally, I want to thank the following individuals who provided pertinent information for this book: Ms. Sarah E. Hines, Executive Director, Alabama Board of Registration for Professional Engineers and Land Surveyors, and Mr. Sherman K. Ward, Programmer/System Analyst, Purdue University.

1

Introduction

THE IMPORTANCE OF TECHNOLOGY

America's economic future depends on its technology base. In fact, the vast majority of U.S. productivity gains during the past half-century, the large number of new industries spawned, and the enormous number of new jobs created are largely due to advances in technology. U.S. companies must be capable of succeeding in world markets in order to ensure that American citizens can maintain a good standard of living in this global economy. Engineers play a fundamental and critical role in the development and exploitation of new technology and are an absolutely necessary ingredient for the nation's future.

The technological changes taking place today are truly phenomenal. No longer is technology constrained to a few industries or found primarily in the laboratories. It is a fundamental part of each person's everyday life. Drastic changes have occurred in the way people live and work over the last several decades, and the rapidity with which these changes are taking place seems to increase every day. These changes are made by individuals who have a sound grounding in math and science. An article entitled "A

Plague of Numbers" that appeared in the November 21, 1991 issue of
Business Week stated the following: "Scientific and mathematical literacy,
and the thinking skills these subjects impart, are no longer a workplace
amenity. They are the hands and backbones of the future work force. A
century ago, a farm or factory could hardly have hired a laborer lacking
quick hands and a strong back. Today, employers can no more afford to
hire workers who lack a quick mind and a strong grounding in science,
math and problem-solving."

As Peter F. Drucker, noted author, management guru, and well-
known consultant, points out in his article entitled "The Age of Social
Transformation" ". . . we live in an era in which for the foreseeable future
it is knowledge, not labor, raw materials or capital that will be the engine
that drives commerce." (See the article in the *Atlantic Monthly,* November
1994, pp. 53–80.) By the year 2000, "knowledge workers," as Drucker
calls them, will make up approximately one-third of the work force in the
United States. This group will be characterized by two important qualifications that did not characterize previous generations:

1. They will require a significant amount of formal education and the
   ability to apply it.
2. They will be continuous learners, constantly updating their knowledge, especially through formal education.

It is important to note, as Drucker points out, that in the '90s the de-
cline in American manufacturing employment has not been caused by im-
ported goods from countries with low labor costs. On the contrary, the
goods (e.g., automobiles) come from countries such as Japan and Ger-
many, where wage rates equal or exceed those in this country. Therefore,
the critical factor involved here is not labor rates but the efficient and effective application of knowledge. Knowledge workers, of which engineers
are certainly one class, will be specialized and, because they are, they will
typically be part of an organization in which perhaps diverse, but comple-
mentary, specialties are employed to solve today's complicated problems.

ENGINEERING TALENT

The Engineering Deans' Council of the American Society for Engineering
Education (ASEE) defines engineering as "the profession in which a
knowledge of advanced mathematics and natural sciences gained by

higher education, experience, and practice is devoted primarily to the creation of new technology for the benefit of humanity." Thus, engineers are problem-solvers who devote their careers to turning concepts and ideas into reality in the sense that they design, manufacture, and maintain processes, products, and services that are economical and safe. As such, the contributions of an engineer extend into a wide variety of areas. The engineer is not only concerned with technology, but the business as a whole, because of the critical role engineers play in a firm's overall performance.

Many people who know little or nothing about engineering tend to classify engineers as weird individuals who spend all their time taking complicated technical gadgets apart to see what makes them tick. Furthermore, they believe that these individuals are shy, reticent, and totally devoid of all social skills, with no interest in the fine arts. This description is absolutely and categorically false. There is no doubt that there are some engineers who fit the mold just defined; however, not many. Through the ages engineers have made significant contributions to society. One notable example is Leonardo da Vinci. Several former presidents were educated as engineers: Presidents Hoover, Eisenhower, and Carter as well as several familiar national figures: Lee Iacocca, Ross Perot, and John Sununu, and a number of the nation's astronauts.

Engineers have a very inquisitive nature. They want to understand the world around them and derive a great deal of pleasure from the mental gymnastics required to solve the complicated problems involved in understanding this world. They are simply motivated by the technical challenge of their work—what author/engineer Samuel Florman calls the "existential pleasures of engineering." Their work is very interesting and quite often they become so wrapped up in it that they think about it while eating, driving, and the like. Contrast this behavior with those individuals whose job is at best just a job and who can't wait to get off work and totally forget about it. To this latter group, engineers are definitely an enigma.

From a talent pool perspective, the current climate in the United States is in a state of flux. According to an article in *Fortune Magazine* (September 21, 1992), the United States has about 1.8 million engineers. Of that number, 92% are male and 89% are white. However, this configuration is in rapid change. The demographics in the United States are changing and therefore there are more opportunities for women and minorities. As a result, academe, government, and industry need and want these underrepresented groups. Furthermore, many of these minorities who are currently in engineering careers have experienced unqualified success. For example, Eleanor Baum is Dean of Engineering at the Cooper Union and Susan Hackwood is the former Dean of Engineering at the University of California at Riverside; Irene Peden is a retired professor of Electrical Engineering at the University of Washington and a former chairman of the Army Science Board; Evangelia Micheli-Tzanakou is Chairman of Biomedical Engineering at Rutgers University; Sheila Widnall is the Secretary of the Air Force; Mae Jemison is the first black female astronaut; Patricia Eng is Senior Transportation Project Officer for the U.S. Nuclear Regulatory Commission; Diana Bendz is Director of Integrated Safety Technology for IBM Corporate Operations and Environmental Affairs; Julie Shimer is Vice President of the Technical Staff and Director of Advanced Custom Technologies at Motorola Inc.; Sherita Ceasar is Director of Manufacturing Operations, Pan American Subscriber Paging Division, Motorola Inc., and Nance Dicciani is the Vice President and Business Director for Petroleum Chemicals at Rohm and Hass. The current position of these women, while in itself impressive, does not tell the entire story. They are involved in a host of professional and other types of activities and have received an impressive list of awards for their outstanding achievements. This is but a small sampling of the enormous success of numerous women in engineering. In addition, similar lists could be prepared for Blacks, Hispanics, and other minority groups. Engineering schools throughout the country are working very diligently to attract women and minorities into the engineering profession.

SOURCES OF INFORMATION

There is a lot of information about the field of engineering, colleges of engineering, and engineering careers that is available for the asking. One of the quickest ways to access this information is through the World Wide Web (WWW). You can use your preferred WWW browser to access what is known as the home page for any organization you seek. As an example, the address for the College of Engineering at Auburn University is http://www.eng.auburn.edu. This address will yield the screen shown in Figure 1.1. If you "click" on the word *electrical,* you will generate the home page for the Electrical Engineering Department, shown in Figure 1.2. If you then "click" on any of the items in this screen, you will obtain all the information that is presented for this topic. Furthermore, you can simply use the WWW browser to search a topic, for example, mechanical engineering. All professional societies that are involved in the accreditation of

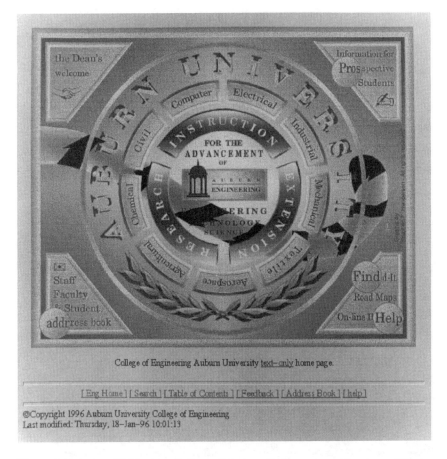

College of Engineering Auburn University text-only home page.

[Eng Home] [Search] [Table of Contents] [Feedback] [Address Book] [help]

©Copyright 1996 Auburn University College of Engineering
Last modified: Thursday, 18–Jan–96 10:01:13

Figure 1.1. Auburn University's College of Engineering home page. (Courtesy of Stephen
F. Henderson)

engineering programs have a web site. For example, the web sites for the
professional societies that correspond to the big four engineering disciplines are

American Institute of Chemical Engineers:
    http://www.che.ufl.edu/aiche
American Society of Civil Engineers: http://www.asce.org
Institute of Electrical and Electronics Engineers:
    http://www.ieee.org/
American Society of Mechanical Engineers: http://www.asme.org/

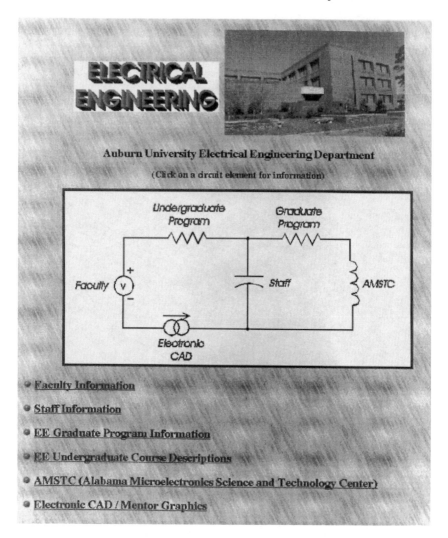

Figure 1.2. The home page for the Department of Electrical Engineering at Auburn University. (Courtesy of Victor P. Nelson)

In addition, these professional societies have either brochures, activity booklets, books, or videotapes that describe their discipline. These societies also have scholarship information—a topic that will be addressed in more detail later in this book. For convenience, all these societies are listed in the appendix. Other sources for career guidance material are

Junior Engineering Technical Society (JETS)
1420 King Street, Suite 405
Alexandria, VA 22314-2715
(703) 548-5387
http://www.asee.org/external/jets

Council on PreCollege Education
American Association of Engineering Societies (AAES)
1111 19th Street, NW, Suite 608
Washington, DC 20036-3690
Tel: (202) 296-2237
Fax: (202) 296-1151
http://www.asee.org/external/aaes

Other organizations that may have material of interest are the National Academy of Engineering (NAE), National Science Foundation (NSF), and the National Institute of Standards and Technology (NIST).

Specific information is also available for minorities and women interested in engineering careers. For example, you may contact these organizations for additional information:

National Action Council for Minorities in Engineering
    (NACME)
3 West 35th Street
New York, NY 10001-2281
(212) 279-2626

Society of Women Engineers (SWE)
345 East 47th Street
New York, NY 10017-2330
(212) 509-9577
http://www.swe.org/

American Indian Science and Engineering Society
    (AISES)
1630 30th Street, Suite 301
Boulder, CO 80301-1014
(303) 939-0023
http://bioc02.uthscsa.edu/aisesnet.html

National Society of Black Engineers (NSBE)
344 Commerce Street
Alexandria, VA 22314-2802
(703) 549-2207
http://www.nsbe.org

Society of Hispanic Professional Engineers (SHPE)
5400 East Olympic Boulevard, Suite 306
Los Angeles, CA 90022-5147
(213) 725-3970
http://www.glue.umd.edu/~shpel

2

Engineering Careers

OVERVIEW

Engineering careers come in many forms, and the work ranges from the very practical to the highly theoretical across a wide spectrum of activities. For example, consider two engineering projects that are at opposite ends of the spectrum—at least in terms of size. One of the largest engineering projects in history is what has come to be known as the "Chunnel." This is the tunnel under the English Channel that links Great Britain with France and is described in Figures 2.1, 2.2, and 2.3. The tunnel spans 38 kilometers under water and is obviously an engineering feat of monumental proportions. Contrast this project with that of the design of a very large scale integrated (VLSI) circuit the size of your little fingernail containing several million interconnected electronic components as described in Figures 2.4 and 2.5. Each of these tasks required numerous engineers with deep knowledge and creative minds.

This chapter focuses on the major career paths in engineering. There is a wide variety of engineering disciplines, which span the entire breadth of technology. The discussion will concentrate here on what is called the

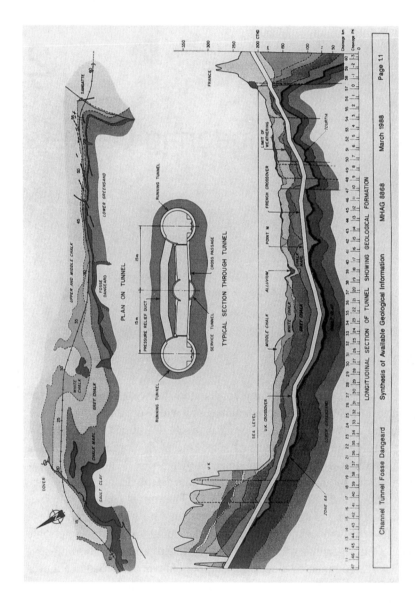

Figure 2.1. Schematic diagrams showing the geological route of the Channel Tunnel under the sea bed between Folkestone and Calais. (Courtesy of QA Photos, Kent, England)

Figure 2.2. One of the four huge (8.72 m) running tunnel boring machines being assembled underground at Shakespeare Cliff. (Courtesy of QA Photos, Kent, England)

big four (in terms of numbers): chemical, civil, electrical, and mechanical. We will also mention other important disciplines and their connection to the big four. This connection is a natural one in a number of cases because many of the other disciplines had their genesis in the big four. Some additional specialties in engineering are aeronautical/aerospace, biomedical, computer, environmental, and industrial.

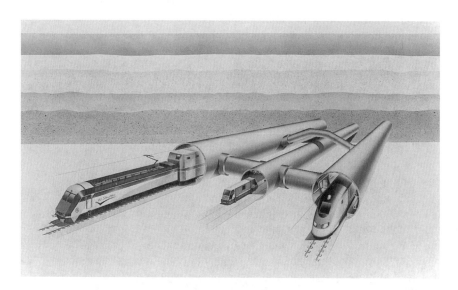

Figure 2.3. Schematic diagram showing the "three-tunnel" design of the Channel Tunnel—two outer running tunnels for Eurotunnel shuttles (left) and Eurostar passenger trains (right) and a smaller service tunnel for maintenance. (Courtesy of

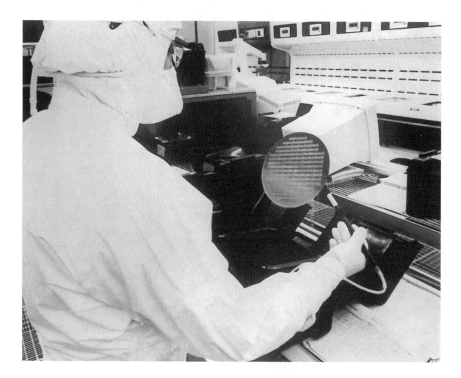

Figure 2.4. A processing engineer examining a 6-in. silicon wafer containing more than 100 68020-microprocessor chips. (Courtesy of Motorola, Inc.)

Figure 2.5. A Motorola 68020 microprocessor chip the size of a fingernail (39.75 mm^2) containing 190,000 transistors dissipating 1.75 watts of power. (Courtesy of Motorola, Inc.) For comparison, the first computer (the ENIAC) was 100 feet long, 8½ feet high, weighed 30 tons, contained 18,000 vacuum tubes and 1500 relays dissipating 150,000 watts. The Motorola chip contains ten times the number of components and is approximately 3,000 times faster than the ENIAC.

Chemical Engineers

Chemical engineers play important roles in a variety of industries in which chemistry is applied in the development of products and processes. The following are some examples of areas of specialization for chemical engineers.

In the biotechnology area, chemical engineers are involved in the successful commercialization of biologically based processes, like high-value pharmaceuticals. Chemical engineers employ advanced biochemical engineering and advanced biotechnology, which result in new products and processes for the benefit of mankind. This area of chemical engineering is particularly suitable for individuals interested in the engineering application of biochemistry, molecular genetics, and applied biology.

In the computer-aided process area, chemical engineers use advanced process simulators on computer workstations, which can effectively model and design advanced manufacturing processes for the chemical, petroleum, plastics, high performance materials, and pharmaceutical industries. They also work on pollution prevention through computer-aided modernization of existing industrial plants and by recovering and recycling unwanted by-products (recycling plastics, solvents, and the like).

In the computer-aided process control area, chemical engineers design and operate advanced control systems for the chemical, plastics, paper, and textile industries. The heart of this technology is the application of advanced computers, workstations, smart controllers, and networks that enable efficient and effective control of chemical, plastics, paper, and textile processes. Chemical process control engineers often work on the application of new process control hardware that has been developed by electrical engineers.

In the field of environmental chemical engineering, chemical engineers work in a wide variety of industries, consulting firms, or government agencies. In these positions, they combine chemical engineering and environmental engineering fundamentals to develop processes that produce a chemically healthy environment. Environmental chemical engineers work to make sure all products and by-products from various chemical-related industries are compatible with the environment through process modifications, waste minimization, advanced waste treatment, environmental control, and so forth.

The paper industry relies on chemical engineers to design environmentally sound and technically advanced paper-making processes as well as developing new methods for recycling and using paper products. Chemical engineers work in paper mills as process and project engineers, applying advanced technology to those mills. Others have the responsibility of production managers. Chemical engineers also work in a variety of companies that serve the paper, plastics, and chemical industries. There are firms that provide engineering services (design, construction, process control, etc.) and firms that provide environmental engineering and chemical services, tailor-making chemical systems that increase product quality or decrease its cost or prevent unwanted pollution. A large number of chemical engineers work for these service companies and equipment manufacturers as product application or technology transfer engineers. These engineers are often teamed up with engineers from the paper mill, plastics plant, or chemical plant, working together to apply their company products or engineering services.

Chemical engineers are involved in advanced energy technologies such as improved petroleum processes and novel energy sources. They develop new forms and sources of liquid and gaseous energy from coal, oil shale, tar sands, biomass, and coal. They also work on perfecting electric car technology by developing advanced electrochemical batteries and capacitors that can give the car a boost of energy for quick acceleration.

Civil Engineers

Civil engineers work on both large- and small-scale projects. They are involved in the planning, design, construction, and maintenance of such structures as buildings, transportation systems, waterways, and a host of large construction projects. They also work on small projects such as city streets, bridges over creeks, town sewers, forest roads, and numerous other activities of these types. They are normally involved in one of the following specialty areas.

In the area of structural engineering, civil engineers design bridges, buildings, offshore platforms, and space platforms. These structures are built with an appropriate combination of such materials as steel, concrete, and timber to resist complicated, dynamic natural forces (e.g., wind and earthquakes) as well as handle their service loads.

Environmental engineering is concerned with protecting people from environmental hazards, disposal of wastes into the environment, and protecting the environment from the effects of human activity. Within this rather broad field, environmental civil engineers design, construct, and manage systems for producing safe drinking water; treat and dispose of wastewater, hazardous materials, and solid wastes; and mitigate surface and subsurface contamination.

Many civil engineering projects are constructed of, or through, rock or soil, and almost all are supported by the earth. Thus, geotechnical engineering affects essentially every aspect of civil engineering work. Civil engineers working in this area design pavements, dams, dikes, and tunnels and foundations for other structures such as buildings, bridges, and offshore platforms. Among the many issues that they must consider are earthquakes and groundwater seepage.

Civil engineers working in hydraulics, hydrology, and water resources engineering deal primarily with the physical processes of water supply and control. Typical work involves agricultural, industrial, and municipal water supply, flood prevention, and the study of propagation of contaminants in groundwater.

Transportation engineering is concerned with the safe and efficient movement of people, goods, and materials. Civil engineers working in this area design, construct, and maintain transportation facilities such as roadways, railways, airfields, and seaports.

In construction engineering, designs are turned into reality in the form of facilities. Civil engineers who work in this environment couple their knowledge in a variety of areas such as financing, planning, and managing, with construction techniques and equipment to successfully build a facility on time and within budget.

Electrical Engineers

Anything that uses electricity in any way, shape, or form is the domain of the electrical engineer. As such, electrical engineers work closely with many other types of engineers on problems that span a number of disciplines. Although the electrical engineering disciplines extend into almost every industry, this field can be subdivided into a number of distinct, but interrelated, areas.

The area of automatic control spans a technology range that extends from aerospace to health care. Within this framework, electrical engineers design and develop automatic landing systems for vehicles such as large airplanes or the space shuttle. They apply control technology in manufacturing (e.g., chemical, paper, or textile) to automatically adjust processes or machinery. Electrical engineers design and develop robots and vision systems for other areas of manufacturing such as automotive and electronics. In the health care industry, electrical engineers design control systems that are used in medical assist devices such as drug injection machines and respirators.

The field of digital systems is tightly coupled to numerous other areas of technology. The reason for this is very simple—the world is going digital: digital control, digital communications, digital electronics, and digital signal processing. Thus, electrical engineers are involved in the design, development, and manufacture of all types of computer equipment, which includes both hardware and software. The equipment ranges from laptops and palmtops to personal computers, engineering workstations, and mini and mainframe computers. Electrical engineers design and develop digital equipment used in communications, such as modems, fax machines,

VCRs, and cellular phones. They embed computers in the control systems of aircraft, nuclear reactors, and manufacturing plants, and employ small microprocessors in video games, home appliances (microwave ovens, refrigerators, and washer/dryers), and in automobiles for ignition control. Digital technology is also the engine that drives the computer-aided design (CAD) systems. Electrical engineers use these CAD systems for the design of VLSI devices and circuits, as well as mechanical components and structures. Electrical engineers use neural networks to emulate the functions of the human brain, and they use high-performance computers in what is known as *virtual reality* to simulate three-dimensional environments. It is important to note that, while this field grew out of electrical engineering, it is also a discipline in its own right and called *computer engineering*.

Electromagnetics is perhaps one of the most highly mathematical areas of technology. At present, it even encompasses optics, because the two areas are merging in frequency. Thus, electrical engineers design and develop radar systems for air traffic control, automatic landing systems, vehicle collision avoidance, terrain mapping, and motion detectors for residential and commercial sites. Another area of increasing importance is the electromagnetic environment. Electrical engineers in this field are concerned with the interaction of electromagnetic fields and biological systems, for example, individuals using cellular telephones.

Electronics, which is now rapidly becoming dominated by microelectronics, is the driving technology behind a vast array of other technological applications that range from one end of the electrotechnology spectrum to the other. For example, it is the foundation for computers, communications—audio, fax, and video—control, and signal processing. Therefore, electrical engineers working in the microelectronics arena are involved in the design, development, and manufacture of smaller, faster, lower power systems, some of which are capable of withstanding extremes in environment, such as radiation and temperature. Furthermore, the electronic instrumentation they develop supports all areas of science, engineering, and medicine.

Electrical power is an industry that affects everyone. Electrical power engineers are primarily concerned with the generation, transmission, and distribution of electrical energy for use in residential, commercial, and industrial applications. Furthermore, they play a similar role for spacecraft because of the nation's goals to explore, live, and work in space. Electrical engineers are also involved in the conservation of energy in manufacturing facilities through computer-controlled energy management systems.

Communications and signal processing are generic to many areas. Electrical engineers working in the telecommunications industry will soon be capable of placing you in instant communication with anyone else in the world. Electrical engineers in this industry have spawned the development of modems, fax, and the rapidly developing cellular technology. In the area of signal processing, electrical engineers design, develop, and manufacture signal processing systems for image, video, geophysical, radar, and sonar applications. One such example is medical imaging for computer automated tomography (CAT) scans, ultrasound, and magnetic resonance imaging (MRI).

Mechanical Engineers

Mechanical engineers work in a wide variety of industries on a diverse set of problems. In large manufacturing facilities, they are concerned with the design, production, and maintenance of all types of dynamic mechanical systems, such as pumps and conveyor belts. They are also involved in the design and manufacture of different types of transportation systems, such as automobiles, trains, and airplanes, as well as farm tractors. Some of the subsystems within automobiles, for example, are fuel-efficient engines, antilock brakes, steering, suspension, transmission, and emission control.

In the area of mechanics, mechanical engineers analyze parts, structures, and systems in order to ensure that the stress and strain levels on the various components are within reasonable levels when they are subjected to static and dynamic loads. The types of elements analyzed range from such things as automobile and airplane parts to tiny electronic packages. This work includes both theoretical and experimental analysis using a variety of sensors and other measurement techniques, and provides the requisite data for the design of components and systems.

In areas such as the power, food, and pulp and paper industries, mechanical engineers design, manufacture, and maintain thermal systems (boilers, water heaters, and heating/air conditioning units). In the electrical/electronic industry, they are involved in cooling electronic equipment, ranging from integrated circuit chips to large power transformers.

Mechanical engineers play a vital role in the research, development, and application of a host of different types of materials, some of which must operate at temperature extremes. These materials include not only such things as metal and plastics, but the new composites for such exotic applications as the construction materials for stealth bombers, materials for appliances in the home (such as Teflon), and the materials used in all types of athletic gear (e.g., skis, golf clubs, tennis rackets, and the like).

The area of acoustic and vibration control is also the domain of the mechanical engineer. Mechanical engineers use both active and passive noise control on all types of rotating machinery, such as diesel engines. In addition, vibration sensors are used in conjunction with electronic equipment as a diagnostic tool for determining the vibration signature of a machine, which, when correlated over time with wear, can be used for maintenance and failure prevention.

Mechanical engineering encompasses a number of other areas of technology. One such area is bio-engineering. In this field, mechanical engineers may team with life scientists to design and build prostheses for arms, knees, legs, and so forth. In addition, mechanical engineers employ computer-aided design tools for the design of mechanical and thermal systems, and develop laser techniques for diagnostics including stress, fluid motion, and thermal changes.

Aerospace Engineers

A discipline that some consider a subset of mechanical engineering is aeronautical/aerospace engineering. Engineers who specialize in this field are normally employed in the aerospace manufacturing industry or government aerospace agencies such as NASA or the FAA. It is perhaps an ideal field for anyone interested in a career in aviation or space exploration. Typical areas of specialization within this discipline are described next.

In aerodynamics, which is the study of the flow of air around and through bodies, engineers are concerned with such things as the analysis and design of aircraft and other vehicles (e.g., cars and trucks) to optimize lift and minimize wind resistance, or drag. Such work is typically performed using a combination of wind tunnel and mathematical techniques, which include computational fluid dynamics and boundary-level theory.

The area of propulsion involves systems that include air-breathing devices like reciprocating and jet engines and liquid-fueled rockets, as well as solid-fueled rockets. These propulsion systems directly impact everything from general and commercial aviation to military aircraft and satellites in orbit.

Structures and structural dynamics are fundamental topics for aerospace engineers. Aerospace structures must not only be flexible and made of lightweight materials, but many of these new composites are smart materials, capable of sensing some parameter and deforming according to some prescribed plan. A study of the dynamics of structures in flow fields involves such things as vibration and the interaction of the structure with the fluid medium, whether it is air or liquid.

Another fundamental topic for aerospace engineers is flight/astro dynamics. This area defines the study of the way aircraft and missiles/space vehicles fly.

Space technology is an area that encompasses quite a large variety of things that significantly impact the nation's standard of living. While space offers a number of advantages, it is a very hostile environment in which to work. It provides a vantage point to work from, unlike anything that can be achieved on earth. Satellites, placed there, are not only used for communication, but for navigation [e.g., the global positioning system (GPS)] and remote sensing. Space has also become a venue for manufacturing in that the zero-gravity environment can be used to produce items like perfect crystals and high-purity drugs.

Biomedical Engineers

Biomedical engineers apply engineering expertise to problems in medicine and biology. These professionals work in an interdisciplinary environment for the enhancement of health care.

One of the areas in which biomedical engineers work is that of bio-instrumentation. In this area, these individuals apply electronics, computers, and measurement techniques to develop devices for the diagnosis and treatment of disease.

In the biomechanics area, they apply the principles of mechanics to develop such devices as artificial hearts, kidneys, and hips.

The area of biomaterials is one in which engineers apply an understanding of both living tissue and materials for the development of implantable materials that will not have adverse effects on the human body and yet are strong enough to last a lifetime.

Clinical engineering is an area in which the biomedical engineer works in a hospital environment to help select the proper equipment for patient care. In addition, engineers ensure that the hospital's equipment is both safe and reliable.

A new and growing area in which biomedical engineers work is that of rehabilitation engineering. In this field, the task is to apply their expertise to a physically impaired individual in order to expand the capabilities of that individual and improve his or her quality of life.

Computer Engineers

As the name implies, computer engineers are involved in essentially every aspect of computing machinery. They design new computer architectures and systems that will be more powerful, faster, efficient, and economical.

They design the interfaces between hardware and software systems, as well as the software that is used to control the various hardware elements in the computer in order to perform specific functions.

Computer engineers play a fundamental role in the design of very large software systems that are used in air traffic control, telecommunications switching, business database management, financial transaction processing, and the like.

They design computer-aided software engineering tools and methods to automate and manage the development of large software projects, and they design computer networks and communication protocols that make possible the efficient sharing and dissemination of information, as in the Internet.

Computer engineers also work on artificial intelligence and design the user-friendly interfaces that make it easier for laypersons to perform tasks on the computer.

Environmental Engineers

Environmental engineers are somewhat unique in that they combine a knowledge of modern technology, biology, and chemistry in order to clean up the environment. The treatment facilities that they design are based primarily on the self-cleansing principles used in nature. These engineers are typically involved in four general areas.

In the area of air pollution, environmental engineers are involved in such things as air sampling and the analysis of inorganic, organic, and particulate pollutants; air quality monitoring; and process and system control.

Environmental engineers design treatment facilities for wastewater, use chemical treatments for the production of potable water, and use a variety of techniques for the control of surface and subsurface contamination.

They are involved in the use of controlled landfill operations for solid wastes and the isolation of toxic chemicals and radioactive materials to prevent these materials from entering the environment.

It is interesting to note that most environmental engineers have a civil engineering background and the next most prevalent background is chemical engineering. Furthermore, civil engineers who work in the environmental area typically do so in either consulting firms or government, while chemical engineers who work as environmental engineers do so in either consulting engineering firms or the chemical process industry.

Industrial Engineers

Industrial engineers are concerned with integrating technology, people, and equipment to obtain optimal system performance. Although initially applied in the manufacturing sector, they now work in widely diverse fields such as transportation, banking, health care, lodging, telecommunications, and military and government organizations. Following are descriptions of some areas of specialization in industrial engineering.

In the occupational safety and ergonomics area, industrial engineers recognize, evaluate, and control hazards that would injure people in the workplace, as well as evaluate and design physical work environments to optimally utilize the capabilities of people while minimizing job-related stress.

The manufacturing, government/military, and service sectors rely on industrial engineers to do operations research and engineering statistical analyses. In this capacity, they support forecasting, resource allocation, scheduling and routing, supply chain logistics, quality assurance and control, design of experiments, and data analysis.

In the area of production and manufacturing systems, industrial engineers are a vital part of designing, controlling, and managing production and distribution systems in virtually every industry. As such, they are involved in manufacturing system operation and design, facility design, production planning and control, warehousing, and inventory management.

Technical marketing and management are other areas where industrial engineers contribute. This area involves not only the obvious fields of marketing and management, but engineering economics, accounting, and finance.

Other Engineering Disciplines

Although the vast majority of engineering careers are those previously discussed, there is a wide spectrum of other engineering specialties:

- Automotive engineering
- Agricultural engineering
- Architectural engineering
- Bio-engineering
- Ceramic engineering
- Fire protection engineering
- Forest engineering
- Geological engineering
- Geothermal engineering
- Heating, Ventilating, and Air Conditioning engineering
- Human Factors Engineering (Ergonomics)
- Manufacturing engineering
- Materials engineering
- Metallurgical engineering
- Mineral and Mining engineering
- Naval engineering
- Nuclear engineering
- Ocean engineering
- Optical engineering
- Petroleum engineering
- Plastics engineering
- Robotics engineering
- Safety engineering
- Software engineering
- Systems engineering
- Textile engineering
- Transportation engineering

Engineers within any one of the disciplines outlined above may work on an individual basis or as a team member in widely diverse areas. For example, some typical titles for engineers that are common to many disciplines are listed below.

- Applications engineer
- Customer engineer
- Construction engineer

- Design engineer
- Development engineer
- Environmental engineer
- Field engineer
- Manufacturing engineer
- Operations engineer
- Project engineer
- Production engineer
- Quality engineer
- Reliability engineer
- Research engineer
- Safety engineer
- Sales engineer
- Test engineer

Engineering careers provide opportunities for a wide variety of creative activities, involvement in interdisciplinary work, and the ability to contribute as both an individual and as a member of a team. It is also important to note that engineers from all disciplines that display leadership potential are often quickly moved into supervisory or management positions. In a normal progression, this movement into management is typically first into technical management and then into general management.

In addition to the areas that are traditionally in the engineering domain, engineers are used in a wide variety of professional areas of business, such as purchasing, cost accounting, servicing, and even law where their unique knowledge and special skills are needed.

Finally, for completeness, it should be noted that, in addition to the engineering career paths outlined above, a number of schools have programs in what is called "Engineering Technology." The Accreditation Board for Engineering and Technology (discussed in Chapter 3) in their Annual Report make the following statement about this area:

> Engineering technology is that part of the technological field which requires the application of scientific and engineering knowledge and methods combined with technical skills in support of engineering activities; it lies in the occupational spectrum between the craftsman and the engineer at the end of the spectrum closest to the engineer. The report further states, engineering problems require solu-

tions of varying degrees of complexity and are constrained by both technical and non-technical considerations. As the technical leader, the engineer determines the policy basic to technical solutions and exercises responsibility to society in the non-technical dimensions. The technician and the technologist work in many functional and responsive ways to execute the applications designed by the engineer.

3

Preparation for Entering an Engineering Program

START IN JUNIOR HIGH SCHOOL

Although some people believe that an interest in engineering and science should be cultivated from grammar school on, junior high school is a good time to start planning to enter an engineering program. Those who know at that time that they want an engineering career are fortunate because they can enter what is often referred to as the "fast track." For example, students would begin taking the proper sequence of mathematics courses in the eighth grade and proceed as follows:

8th grade—Algebra I
9th grade—Algebra II
10th grade—Geometry
11th grade—Trigonometry
12th grade—Advanced Placement (AP) Calculus with credit

The approach listed above for mathematics can also be followed for course sequences such as Chemistry, Physics, English, History, Biology, and the like. This approach is called the fast track because if the courses in high school are of sufficient quality, the students may be given some credit for these courses in college and thus may skip ahead to the more advanced material. At the very least, however, they will have the necessary fundamentals to enable them to get the most out of their engineering program.

Many students do not decide early that they are interested in an engineering career. Those individuals hopefully will have finished trigonometry in high school and therefore can enter the university and begin taking calculus. If trigonometry has not been completed in high school, noncredit remedial or supplemental work must be taken before the proper mathematics sequences can be started. This remedial or supplemental work, if necessary, can be taken at community colleges that are typically close to home. Check with your high school guidance counselor for locations.

THE ACT AND SAT

The ACT and SAT are standard tests used as a gate by universities in their admission requirements. The ACT is the American College Test and the SAT is the Scholastic Assessment Test. The ACT is developed by American College Testing in Iowa City, Iowa. The SAT is developed by Educational Testing Service in Princeton, New Jersey. Some universities use one test, some use the other, and some use both. The ACT score ranges from 1 to 36 and the SAT score ranges from 400 to 1600. The SAT was recently revised in order to recenter the scores, and the stated equivalencies between the SAT and the ACT are shown in Table 3.1.

Table 3.1 ACT/SAT Equivalent Scores

ACT / SAT		ACT / SAT		ACT / SAT		ACT / SAT	
1	400	10	430	19	910	28	1240
2	400	11	480	20	950	29	1280
3	400	12	540	21	990	30	1320
4	400	13	600	22	1030	31	1360
5	400	14	670	23	1060	32	1410
6	400	15	720	24	1100	33	1460
7	400	16	770	25	1140	34	1530
8	400	17	820	26	1170	35	1580
9	400	18	860	27	1210	36	1600

It is interesting to note that the two tests seem to test slightly different things. For example, an individual born with a high IQ will score high on the SAT, while an individual who may not have a high IQ but is a hard-working individual may score well on the ACT.

Students should take the ACT or SAT as juniors in high school. If the test is taken early in their junior year, they have a chance to correct any deficiencies that the test may illuminate.

It is impossible to overemphasize the importance of doing well on these standardized tests. In support of this position, consider the following data obtained by examining the records of freshmen entering the College of Engineering at Auburn University. Because Auburn ranked tenth in the nation in the production of engineers in 1993 and approximately 35 percent of the students are from out-of-state, it would appear that the data represents a reasonable sample of the engineering freshman class nationwide.

The engineering freshman class that entered Auburn in the Fall of 1991 was tracked for two academic years, which is the maximum time that they can remain in pre-engineering (i.e., as a freshman) before transferring to a major such as Civil, Chemical, and so on. In order to transfer to a major, students must have a minimum grade point average (GPA—a term discussed in some detail later) of 2.2 on a 4.0 system. Simply stated, the students' overall grades must be better than a C average.

There were 672 students with ACT scores in the sample. Table 3.2 illustrates the correlation between ACT scores and successful admission into an engineering major. For example, of the 12 students with an ACT score of 18, only 25 percent, or 3 students, were admitted to engineering. The remaining nine students left the engineering college for various reasons, but primarily due to academic difficulty. The data clearly indicate that the higher the ACT score, the greater the probability of success. The academic preparation of the students at the low end of the scale is simply inadequate to be able to perform at the level required. However, it is equally important to note that high ACT scores are not a guarantee of success in an engineering program. Even these more capable students need to develop the skills addressed later in this book.

Personnel in Purdue University's Department of Freshman Engineering have studied success indicators for engineering freshmen for almost three decades. They have found that there are reliable indicators in addition to the ACT or SAT score. Prior to entering the university, the best success indicators for students are provided by their performance in specific high school mathematics courses (algebra, trigonometry, solid geometry, and precalculus) and the science courses (e.g., chemistry and physics).

TABLE 3.2   ACT Scores of the Freshman Class that Entered the College
of Engineering, Auburn University, in the Fall of 1991

ACT score	Successfully admitted into engineering		Unsuccessful due primarily to academic difficulty		Total number
	Number	Percent	Number	Percent	
16			1	100	1
17			4	100	4
18	3	25	9	75	12
19	8	25	24	75	32
20	10	27.8	26	72.2	36
21	16	41	23	59	39
22	15	32.6	31	67.4	46
23	28	50.9	27	49.1	55
24	31	44.9	38	55.1	69
25	27	40.9	39	59.1	66
26	37	56.1	29	43.9	66
27	37	63.8	21	36.2	58
28	38	70.4	16	29.6	54
29	31	63.3	18	36.7	49
30	28	73.7	10	26.3	38
31	22	88	3	12	25
32	10	76.9	3	23.1	13
33	5	83.3	1	16.7	6
34	3	100			3
Total	349	51.9	323	48.1	672

Once a student is accepted by the university, the best success indicators are
the grades obtained during their first semester at the university.

SELECTING A UNIVERSITY

There are two sources that provide a considerable amount of information
about engineering schools across the country. They are the American Society for Engineering Education (ASEE) and the Accreditation Board for
Engineering and Technology (ABET). The addresses of both these organizations are given in the appendix.

An item that naturally arises in conversations with admissions personnel and other university officials is the ranking of the university. The

universities in the United States are typically ranked periodically and the information can be found in such publications as *U.S. News and World Report*. Furthermore, *U.S. News* publishes a guidebook that lists America's best colleges and offers a college planner. The universities that are ranked at the top of the list are truly outstanding institutions of higher learning. However, it is important to keep in mind that this ranking is based on such things as the number of faculty who are members of the National Academy of Engineering, the amount of research that is performed by the faculty, the number of papers that the faculty have published in peer-reviewed journals, the number of Ph.D.'s that are produced at the institution, and the like. (Although it is implied in many ways in the items just listed, conspicuous by its absence is the word *teaching*.) A young man or woman entering a university should certainly be concerned about the quality of the program that they are about to enter, but there are many schools that have something to offer that are not in the top of the national ranking list. For example, important questions that should be asked include:

- Is the faculty competent and dedicated to providing a quality education for their students?
- Are they available to talk with students about their course work or other significant matters?
- What are the class sizes?
- Will the classes be primarily taught by graduate students?
- What is the general atmosphere of the campus?
- Will your car be stolen while you visit the admissions office?
- Will your coat disappear while you return your tray in the student center cafeteria?

These questions and a host of other similar questions that have a direct bearing on the quality of life, as well as education at the institution, should be asked.

While there are numerous and varied reasons for selecting a particular university, some of the more prevalent ones are the strength of the academic programs, the costs (which also involve financial aid), job prospects for graduates of the school, the proximity of the university to the student's home, and the availability of well-balanced social activities. Regardless of the specific criteria used in your selection of a university, it is wise to select more than one and prioritize them in your own mind. It is unfortunate, but the reality of this situation is that you may not be accepted by your first choice. Thus a backup plan helps eliminate delays and problems at a point when time could be crucial.

VISITS TO CAMPUS

Having selected a number of potential universities, it is important to visit each one. Visits to the university campus should take place in the summer between the student's junior and senior years, because the applications should be made to the university in the August/September time frame. Prospective freshmen are encouraged to visit at least two schools in order to have a basis for some type of comparison. One full day during the normal work week when school is in session should be devoted to the visit so that students and their parents can get a feel for the atmosphere and have sufficient time to look carefully at all the points of interest. These visits should be treated like any other professional appointment and can be arranged by simply calling the admissions office.

A typical campus visit will involve a talk with the admission personnel, where all information concerning the admission standards and requirements is discussed. The visit will also include sessions with personnel who handle housing and student financial aid. Usually, a student currently enrolled in the university or another individual who is well acquainted with the campus will provide a tour. There is also the opportunity to visit the academic area in which the student plans to enroll—for example, the electrical engineering department and/or the pre-engineering advisor.

A campus visit is an excellent opportunity to check out every aspect of the university. Although you may have a number of questions that are directly related to your own personal situation, some of the more general questions that are normally of interest involve costs (including financial aid), housing, class sizes, campus safety, statistics on retention, and graduation rates.

ADMISSION

Once the various potential universities have been selected and visited, it is now time to apply to all those that, at this point, are viable. It is wise to begin the admission process sometime between the junior and senior year in high school, but no later than the start of the senior year. By this time, the ACT or SAT scores are available. This seemingly early start provides sufficient time to apply for scholarships and other financial aid. Some students want to live in dormitories on the campus, and these will fill quickly. Some state universities do not discriminate against out-of-state students, but they may do it late in the game when the entering freshman class begins to reach its maximum level. In addition, if any problems surface, there is time to work them out if the application is received early. For example, if the student's grades are marginal, there may be time to take the ACT or SAT again, or take additional work at a junior college. Furthermore, many universities do not have quotas, but if one does, the earlier the application is received, the better.

The admission requirements vary from university to university, but are always documented in what is called the university's catalog or bulletin. Typically, the admission requirements are based on the ACT or SAT and the high school GPA, provided that the high school courses meet core class requirements in mathematics, English, and the like.

Finally, the enrollments in engineering have remained stable and relatively constant for many years. The number of full-time students entering U.S. engineering programs each year is normally a little more than 90,000. The total full-time enrollment in engineering programs is approximately 330,000. This total figure includes about 18 percent women and 14 percent underrepresented minorities.

THE TIMETABLE

To summarize some of the time schedules mentioned earlier, the process of preparation for entering a university to study engineering should be undertaken in the manner outlined in Figure 3.1. The normal track for the mathematics and science courses to be taken each year is shown in the solid block diagram. High schools that offer more advanced courses permit the students to develop an accelerated program that culminates in calculus and advanced courses in science, as indicated in the dashed-line diagram. Students are advised to look at the terminal course at their school (i.e., precalculus, calculus, or whatever) and work backwards from that point to identify their plan of study. Keep in mind that what the school hopes to have in the future may or may not become reality. Furthermore, if a student is not on either the fast or normal track, all is not lost. The student should simply visit the high school or college advisor who will work with him or her to develop a viable plan.

The PSAT, or Pre-SAT, can be taken in the sophomore year of high school. This test is typically used for identifying National Merit Scholars.

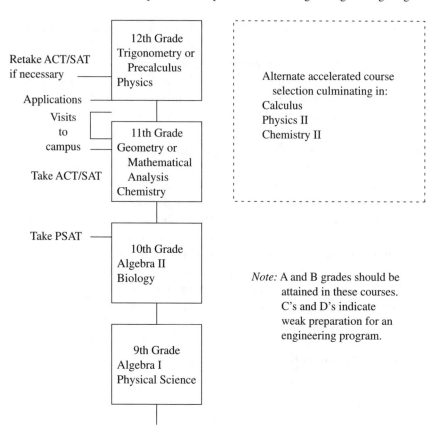

Figure 3.1. Timetable for preparing for an engineering program.

The ACT or SAT should be taken in the spring of the junior year of high school. The visit to the campus should be done in the summer between the junior and senior years of high school. The application to the university should be submitted at the end of the high school junior year, and this application will automatically trigger a number of other items such as housing, financial aid, and scholarships. Finally, many schools have a pre-college counseling session or orientation program in the summer prior to fall enrollment, and prospective students should attend this first-time university experience.

There are numerous advantages to following the suggested timetable. For example, if you take the ACT or SAT in the spring of the junior year, you can take it again in the fall of the senior year. Taking a test such as this is a learning experience, 36 and many schools will take the highest score. This timetable also gives you time to think carefully about all your options so that you select the best college from among your choices. Furthermore, at many schools you cannot apply for campus housing or be awarded a scholarship until you are accepted as a student.

COMMUNITY/JUNIOR COLLEGE TRANSFER

Some students, for any number of reasons, do not go directly to the university, but rather enter a community or junior college with the intent of completing their first two years and transferring to the university at the beginning of their junior year. This approach, like any other, requires planning. It is important, at the outset, that you apply early and check carefully the interrelationship that exists between your junior college (JC) and the university to which you ultimately desire to transfer. One of the first questions you want to ask is, will the courses that you take at the JC transfer directly to the university. If the answer to that question is satisfactory, then you want to know if the JC will offer the courses you need to transfer to the university in a timely fashion. If you are prepared and do not have to take remedial work, you should be able to take at the JC the calculus, chemistry, physics, English, and other humanistic/social science courses (psychology, sociology, etc.) that are required in a university engineering program. You must typically maintain a B or better average overall with a minimum of a B average in the mathematics and science courses in order to demonstrate the qualifications for success in an engineering program. The minimum standards for course transfer credit are actually less, as discussed later. There are a series of courses in mathematics, chemistry, and physics that must be taken at some point. If you complete the series in mathematics (e.g., Math I, II, and III) at the JC, then there is no problem with transfer credit, because the entire series constitutes the requirement. However, if you complete only a portion of the courses at the JC and wish to transfer to the university, there is a potential problem because there could be a slight mismatch between the university math sequence and that offered at the JC. For example, such a mismatch could result if the JC is on the semester system and the university is on the quarter system. By taking the required sequence at the JC, you may exceed the hour requirement for the courses, but you will not encounter transfer problems and you will have a solid foundation upon which to make more intelligent decisions about your degree path when you transfer to the university. Long-term planning and a pretransfer visit to see the transfer advisor at the engineering college of your choice will eliminate potential problems. (Use this opportunity to apply the problem-solving skills you will need as a future engineer.)

4

Engineering Curricula

FUNDAMENTAL CONCEPTS

The Engineering Deans' Council of the American Society for Engineering Education (ASEE) states that "engineering education focuses primarily on the conceptual and theoretical aspects of science and engineering aimed at preparing graduates for practice in that portion of the technological spectrum closest to the research, development, and conceptual design functions." Whatever else is said about an engineering education, one thing is for sure—it educates you to understand the world around you and teaches you to contribute to it so that you can enhance society's standard of living.

The world is becoming more sophisticated every day, and advances in technology have fostered developments that have significantly changed the way in which people live. Only a moment's reflection is necessary to catalog a number of new technological developments that have had a staggering effect on lives. As a single example, consider the developments in modern electronics. This ubiquitous technology has drastically changed numerous products used everyday, such as automobiles, home appliances, radios, televisions, telephones, fax machines, personal computers, and VCRs. Because of the phenomenal effect that technology has had on the lives of every citizen and the profound changes that are going to take place in the future, *an engineering program is actually the liberal arts education of the twenty-first century.*

ACCREDITATION

An accredited engineering program at a university gives assurance to potential students and their prospective employers, that high-quality engineering education is provided at that university.

The sole agency responsible for the accreditation of programs leading to degrees in engineering is the Accreditation Board for Engineering and Technology (ABET).

In general, ABET establishes standards and goals designed to enhance the practice of the profession and promotes the intellectual development of those individuals who are interested in engineering and the related professions. The accreditation process for engineering programs is carried out through the Engineering Accreditation Commission (EAC), which is part of ABET.

Recent data indicate that within the United States there are 315 institutions with 1,516 accredited engineering programs. Including the nontraditional program area, there are 25 different program areas defined for accreditation. The following is an alphabetical listing of these programs.

Aerospace Engineering
Agricultural Engineering
Bio-engineering
Ceramic Engineering
Chemical Engineering
Civil Engineering
Computer Engineering
Construction Engineering
Electrical Engineering
Engineering Management
Engineering Mechanics
Environmental Engineering
Geological Engineering
Industrial Engineering
Manufacturing Engineering

Materials Engineering
Mechanical Engineering
Metallurgical Engineering
Mining Engineering
Naval Architecture & Marine Engineering
Nuclear Engineering
Ocean Engineering
Petroleum Engineering
Surveying Engineering
Nontraditional Programs

The ABET is a confederation of 26 participating bodies and 6 affiliate bodies. These participating bodies are the engineering professional societies, such as the American Institute of Chemical Engineers, the American Society of Civil Engineers, the American Society of Mechanical Engineers, the Institute of Electrical and Electronics Engineers, and the like. An example of an affiliate body is the American Consulting Engineers Council.

The criteria for evaluating programs in engineering is divided into two categories: general criteria and program criteria. The general criteria are designed to assure that the students receive the proper foundation in mathematics, basic science, engineering sciences, engineering design, and humanities and social sciences; and, in addition, the proper preparation in a higher engineering specialization. The program criteria are developed by the cognizant participating bodies in conjunction with the ABET Board. For example, the American Institute of Chemical Engineers will develop criteria for programs in chemical engineering, subject to approval by the ABET Board. The general criteria are concerned with the following items: faculty, the curricular objectives and content, the student body, the administration, the institutional facilities, and the commitment that the institution makes to the program.

The very heart of any educational program is the faculty and, therefore, in the list of items covered in the general criteria, they are listed first. The criteria is concerned with their competence, numbers, composition, teaching and research loads, and generally anything that affects their ability to provide a solid education for their students.

The curricular objectives are set by each school and designed in such a way that they develop in the student an ability to apply pertinent knowledge to the solution of engineering problems in both an effective and professional manner. While two similar programs in, let's say, civil engineering may be accredited, their curricular objectives may be different. By examining these curricular objectives, students can match their career goals to the objectives of the particular program. For example, a civil engineering student who wishes to pursue a career in environmental work may select a program different from another individual who is interested in a career in bridge design.

The curricular content for programs in institutions that prepare graduates to enter the engineering profession upon graduation with a bachelor's degree, must provide an overall integrated educational experience. Each curriculum consists of a carefully selected set of requirements that are designed to take an entering freshman through a systematic program in a very methodical progression of topics to produce an engineer in the specific discipline. Studying to become an engineer is a very interesting and rewarding experience, and it is fun to learn the technology and its ramifications.

Of the normal four years that are required to obtain an engineering degree, three of the years are specified in the following manner:

- One year of an appropriate combination of mathematics (beyond trigonometry) and basic sciences. The mathematics courses must include differential and integral calculus, differential equations, and perhaps additional work in one or more of the following subjects: probability and statistics, linear algebra, numerical analysis, and advanced calculus. The basic science courses are typically chemistry and physics, with perhaps additional course work in life sciences, earth sciences, or advanced courses in chemistry or physics.

- One and one-half years of an appropriate combination of engineering topics. Engineering topics include both engineering science and engineering design. The engineering science courses include such topics as mechanics, thermodynamics, electrical and electronic circuits, material science, transport phenomena, and computer science. Engineering design topics are integrated throughout the curriculum. Specific design courses normally appear in the upper-division level of a program and include such things as formulation of design problem statements and specifications, use of design methodology, consideration of alternate solutions, as well as realistic constraints (e.g., economics, reliability, and safety).

- One-half year of humanities and social sciences. Traditional courses in these areas are anthropology, fine arts, foreign language, history, literature, political science, psychology, religion, sociology, and the like.

Some fairly typical examples of semester-based, as opposed to quarter-based, curricula for chemical, civil, electrical, and mechanical engineering are shown in Tables 4.1 to 4.4, where the credit hours for the various courses match the ABET requirements. In addition, Figures 4.1 to 4.4 illustrate some of the laboratory work that students in these four curricula may take in the latter part of their program.

Additional factors considered in the decision to accredit a program include the following: an appropriate laboratory experience, the use of computers, probability and statistics in problem-solving activities, competency in written and oral communication, and an understanding of ethical, social, economic, and safety considerations.

The quality and performance of the student body, while in school as well as after graduation, is an important consideration. When the students are well prepared, the level and pace of their instruction can be high.

The administration of the college of engineering, through effective and efficient leadership, plays a key role in enhancing the ability of the faculty to provide proper instruction. The attitude and policies of this administration provide the atmosphere under which all instructional activities are executed.

Institutional facilities, such as classrooms and laboratories, the library, computing equipment, and the like, are very necessary for the support of a quality education in engineering.

Finally, the institution's commitment to the engineering programs, both financially and philosophically, is essential. In order for engineering programs to achieve their educational objectives, the central administration must ensure that there are sufficient numbers of faculty and that they have the resources necessary to provide the proper instruction.

The program criteria, which are submitted by the cognizant participating bodies, typically delineate in some detail, specific requirements concerning the faculty, curriculum, and facilities. For example, the program criteria may list certain types of computer software that must be an integral part of the program.

The requirements outlined above are not a hurdle, but rather a collection of items that ensure, for the student, a pathway to an exciting career.

The details surrounding the items listed above, as well as all the facets and ramifications of the engineering accreditation process, can be

Table 4.1 A Typical Semester-Based Chemical Engineering Curriculum

Process Design 2	Technical Elective	Technical Elective	Fine Arts Elective	Free Elective	Senior Level
Process Design 1	Chemical Reactor Design	Computer Applications in Chemical Engineering	Technical Writing	Chemical Engineering Elective	
Mass Transfer 2	Process Control	Hazardous Material Management	Fluid Mechanics	Chemical Engineering Laboratory	
Mass Transfer 1	Circuits	Heat Transfer	Engineering Mechanics 1	Physical Chemistry	
Math 4	Physics 2	Literature 2	Thermodynamics 2	Organic Chemistry 2	
Math 3	Physics 1	Literature 1	Thermodynamics 1	Organic Chemistry 1	
Math 2	Ethics	Public Speaking	History 2	Basic Chemistry 2	
Math 1	Computer Programming	English Composition	History 1	Basic Chemistry 1	Freshman Level

Table 4.2 A Typical Semester-Based Civil Engineering Curriculum

Design Project	Technical Elective	Technical Elective	Fine Arts Elective	Construction Management	Senior Level
Concrete Design	Wastewater Treatment	Steel Design	Technical Writing	Engineering Economics	
Water Treatment	Statistics	Foundations	Fluid Mechanics	Structural Analysis 2	
Wastewater Collection/Water Distribution	Highway Design	Soil Mechanics	Engineering Dynamics	Structural Analysis 1	
Math 4	Geology	Literature 2	Strength of Materials	Surveying	
Math 3	Circuits	Literature 1	Engineering Statics	Physics 2	
Math 2	Ethics	Public Speaking	History 2	Physics 1	
Math 1	Computer Programming	English Composition	History 1	Chemistry 1	Freshman Level

Table 4.3 A Typical Semester-Based Electrical Engineering Curriculum

Design Project	Technical Elective	Technical Elective	Fine Arts Elective	Free Elective	Senior Level
Modern Physics	Technical Elective	Communications / Communications Lab	Technical Writing	Engineering Economics	
Electronics 2 / Electronics Lab	Probability/Statistics	Electromagnetics 2 / Electromagnetics Lab	Control Systems / Control Lab	Engineering Science	
Electronics 1	Linear Systems	Electromagnetics 1	Electric Power / Power Lab	Material Science	
Math 4	Circuits 2 / Circuits Lab	Literature 2	Digital Systems 2	Math 5	
Math 3	Circuits 1	Literature 1	Digital Systems 1	Physics 2 / Physics Lab	
Math 2	Ethics	Public Speaking	History 2	Physics 1 / Physics Lab	
Math 1	Computer Programming	English Composition	History 1	Chemistry / Chemistry Lab	Freshman Level

Table 4.4 A Typical Semester-Based Mechanical Engineering Curriculum

Design Project 2	Technical Elective	Technical Elective	Fine Arts Elective	Free Elective	Senior Level
Design Project 1	Technical Elective	Free Elective	Technical Writing	Engineering Economics	
Machine Design	Mechanical Engineering Design	Dynamics of Machines	Control Systems	Engineering Dynamics	
Circuits	Fluid Mechanics	Material Science	Design for Manufacturing	Heat Transfer	
Math 4	Thermodynamics 2	Math 5	Strength of Materials	Math 6	
Math 3	Thermodynamics 1	Literature 2	Engineering Statics	Physics	
Math 2	Ethics	Literature 1	History 2	Physics 1	
Math 1	Computer Programming	English Composition	History 1	Chemistry	Freshman Level

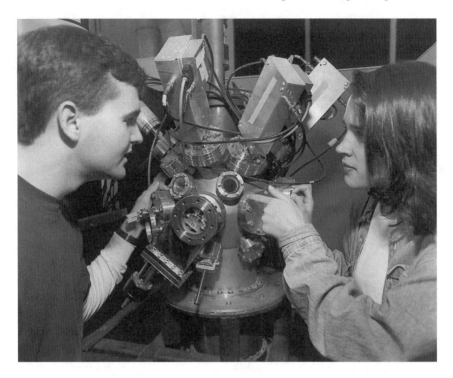

Figure 4.1. Students performing a surface analysis using a spectrometer.

found in the ABET Annual Report. This document is published by ABET and can be obtained at the following address:

> Accreditation Board for Engineering and Technology, Inc.
> 111 Market Place, Suite 1050
> Baltimore, MD 21202-4012
> Telephone: (410) 347-7700

STUDY HABITS

Study habits will be a key success factor in an engineering program. It would appear that some of the young men and women entering the university today are immature and lack a sense of responsibility. They are unlike those who understand discipline, hard work, and sacrifice. If students

Figure 4.2. A student preparing an asphalt sample for a dynamic creep test.

are bright and on the fast track in high school, they probably will make good grades without really studying hard. Because they did well in high school, their first term at the university can be relatively easy. However, the material studied in high school is covered quickly, and during the second term the material gets more sophisticated and intense. At this point, the student begins to realize that taking engineering is like taking a sip of water from a fire hose. One must then learn to cope. It is important that students determine the level at which they can effectively operate and understand the proper balance between class loads and other activities that must be achieved in order to survive or, better, to flourish.

One's social life and extracurricular activities must be placed in proper perspective. Doing well in your classes must come first and, as a general rule, that means studying must be given the highest priority. It is important, however, to keep a balance of social life in order to maintain a positive and progressive attitude. Social maturity in relating to other people is a necessary ingredient in essentially every engineering career.

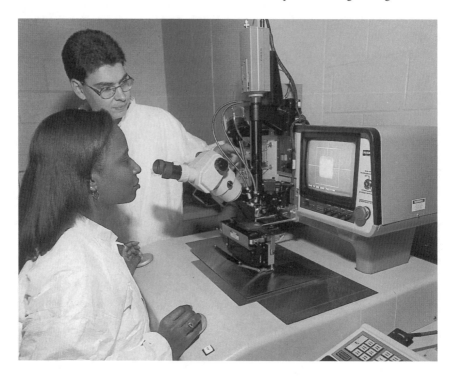

Figure 4.3. Students performing a wirebonding operation for an integrated circuit chip.

It is very important that you have a quiet place in which to study. If your apartment or dormitory does not provide the proper setting, then find a place that will. For example, the university library is always available, and there are often rooms in selected buildings that are set aside as study halls. Furthermore, it is very helpful to have a study partner who reinforces your commitment to spend the necessary time preparing your lessons. If you have an appointment to meet someone for the purpose of studying, you are less likely to end up in some other activity. For most of us, it is essentially impossible to study while there is a bull session going on in the immediate or adjacent area, the walls are vibrating to the tune of a stereo in the next room, or a television set is blasting away close by. However, if you have a place where you can think and concentrate, you will not only learn better, you will learn quicker and easier.

Students must determine just how much academic horsepower they possess. The best approach is to be conservative early in the game. Engi-

Figure 4.4. Students checking dimensions on a new mini-baha car.

neering students normally find out quickly that they cannot match hour for hour the time that some of their fellow students spend in extracurricular activities, and the quicker they learn it, the better.

Universities are well aware of the problems that students encounter and, as a result, have programs to aid them in coping with the situations in which they find themselves. There is normally an office of counseling services that offers numerous opportunities for help. For example, in addition to individual counseling on such items as study skills and academic development, there are seminars on time management, note-taking, exam preparation, and the like.

The engineering college may even sponsor free, one-on-one tutorial programs where students can get individual attention on any course in an engineering curriculum. This assistance can be especially valuable in helping with some course for which the student seems to have little or no aptitude.

If parents are investing their money in a college education for a son or daughter, it is wise to keep an eye on that investment. Report cards are sent to the student (not the parent) at the end of every term. Parents who are not kept abreast of their children's progress should inquire after each term. When parents do not see a report card, more often than not there are academic problems. The performance data are, however, easily obtained. A letter signed by the student and addressed to the registrar of the university that says "My name is Connie Coed, my social security number is 123-45-6789, and I hereby request that you send a copy of my transcript to the following address . . ." will produce the student's records on a term-by-term basis. Finally, the engineering professors maintain office hours during which students can receive one-on-one help with technical questions. Sometimes it is simply a matter of looking at an issue from a different perspective, and your professors are generally very capable of showing you ways to examine an issue that you may not have considered yourself. Whatever the issue, they have set aside time to do what they can to enhance your learning experience.

COURSE REGISTRATION/SCHEDULES

Ideally, it would be nice to have your lecture courses in the morning hours and your laboratories in the afternoon. However, early in your collegiate career, your schedule may be less than ideal and you may be forced to take a load lighter than you have planned because of the lack of available courses or the number of seats in the classes you want. However, regardless of the availability of classes, it is recommended that during your first one or two terms you consider taking fewer classes than the university catalog recommends. For many individuals, this is not necessary or even a good idea. However, the jump from high school to the university is a big one, and if you are not mentally and psychologically prepared, you may find that a lighter initial load will help make the transition easier. If you get off to a shaky start initially, part of your time at the university will be spent trying to catch up to your real potential. On the other hand, if you get off to a good start, you can probably make up any lost time later.

Course scheduling is much easier when you reach the upper-division courses. During this period, you will typically be able to get at least most of the courses you want, when you want them. The university will try to schedule the classes in such a way that you are able to get what you want, but it is hard to always please everyone. Just remember that you are only one individual in what seems like an endless stream of people who have run this gauntlet and emerged as happy and successful alumni.

The first year of an engineering program will typically contain few, if any, engineering courses. The first year is normally spent taking courses in mathematics, chemistry, physics, and courses that are of a humanistic/social science nature. It is important to understand the content of the first year, because it is the mathematics and science courses that provide the foundation for the exciting courses in engineering. The humanities courses are an important part of the curriculum because they help students develop into solid citizens, who will exhibit high moral and ethical standards in every facet of their lives. It is unfortunate that the engineering courses typically do not begin until the sophomore year. However, some of the changes that are taking place in engineering education today are specifically designed to address this issue. Some schools are now putting courses in the freshman year that introduce the students to engineering topics in a fun and fascinating way. There are other schools that have been doing something like this for some time.

There is another important point that should be understood about the freshman year. It is during this year that students who have not firmly decided on their engineering major make up their minds. In other words, are they going into chemical, civil, or what. Therefore, for those individuals who are not sure of their major, this year is one in which they can study the various areas at close range by visiting the departments and talking to students, professors, and counselors to determine which curriculum best fits their career aspirations.

CLASS ATTENDANCE

With the cost of a university education continuously on the rise, the issue of class attendance would seem to be a moot point. However, there are some students who treat class attendance in a casual manner. In many classes, attendance is not mandatory; therefore, you must be disciplined to attend every one of them. You should feel morally obligated to do the very best you can if your parents or relatives are paying for your education (if you are paying for your own education, you will be present at every class), and you can't possibly do your best if you don't faithfully attend your classes. From a strictly pragmatic viewpoint, if you miss class, you may

miss some critical information, a quiz may be announced, you might even
have a pop quiz that day, assignments may be made for delivery late in the
term and not mentioned again until the last minute, or at the very least the
foundation for understanding subsequent material may be covered in detail that day and not addressed, per se, again.

GRADE POINT AVERAGE (GPA) REQUIREMENTS

Your GPA is a subject that should be near and dear to your heart. In the
university, your GPA is your ticket to adventure. First of all, many schools
clearly distinguish between pre-engineering (the first year or two) and engineering (the remaining two to three years), and you will need a certain
GPA in order to transfer from pre-engineering to engineering. Failure to
achieve the proper GPA will result in a "pink slip" from the college of engineering.

Most schools employ the 4.0 grading system. Therefore, the relationship between grades and quality points/credit hour is as shown in
Table 4.5.

Table 4.5   Conversion Between Grades and Quality Points/Credit Hour

Grades	Quality Points/Credit Hour
A	4
B	3
C	2
D	1
F	0

Now suppose that the university is on this 4 point system and requires 120
semester hours for graduation (there are two terms per year and 15 credit
hours per term). The freshman year is classified as pre-engineering, and
the sophomore through senior years are engineering. In addition, suppose
that a 2.2 GPA is required to transfer from pre-engineering to engineering
and a 2.0 GPA is required for graduation.

Consider the academic record of Connie Coed. Connie came to the
university sure that she wanted to be a mechanical engineer. During her
first term, she took 15 hours of classwork as outlined in Table 4.6 and received the grades listed.

TABLE 4.6 An Example First-Term Grade Report

Course	Credit hours	Grade	Quality points
Math 1	3	A	12
Chem 1	3	B	9
English 1	3	C	6
Hum/Soc 1	3	C	6
Intro/Engr	3	B	9

Connie's GPA for this term was

$$\text{GPA} = \text{Total quality points/number of hours attempted}$$
$$= 42/15$$
$$= 2.80$$

Connie is off to a good start. She must achieve a 2.2 GPA on the 30 credit hours in the freshman year, and she already has a 2.8 GPA with only 15 credit hours remaining.

During the second term, Connie missed some classes as a result of a broken leg, which she received playing intramural soccer. However, she took the classes outlined in Table 4.7 and received the grades shown.

Table 4.7 An Example Second-Term Grade Report

Course	Credit hours	Grade	Quality points
Math 2	3	B	9
Physics 1	3	B	9
English 2	3	D	3
Hum/Soc 2	3	C	6
Comp. Sci	3	B	9

Her GPA for this term was

$$\text{GPA} = 36/15$$
$$= 2.4$$

Connie is to be congratulated on a job well done. With these grades, she has made the cut and will be transferred into engineering. The reason is this: Connie needed a 2.2 GPA on the hours attempted, and because she has attempted 30 hours, she needs $30 \times 2.2 = 66$ quality points. However, during the two terms in school, she has obtained 78. Furthermore, her overall GPA at this point is a 2.6.

While you are classified as an engineering student, your GPA will determine your eligibility for loans, scholarships, jobs, and induction into a number of scholastic honor societies. The higher your GPA, the more doors that will open before you. Students with a high GPA can be asked to join one or more honor societies, and your membership in them becomes a part of your permanent record at the university.

Finally, your GPA will play a big part in your initial future following graduation. If you have a high GPA, you will be sought after by many companies (provided the job market is good), and you will be eligible to go to graduate school, if you wish.

There are those individuals who argue that GPA is a poor measure of performance and not a good indicator of success. However, the reality of this situation is that, like it or not, it is often the key measure used by companies in hiring employees, particularly those graduating from the university in recent years.

TRANSFER CREDIT

Pre-Engineering Courses
for Engineering Majors

If you plan to take the first year or two of engineering at a junior college or small four-year institution, perhaps close to home, and then transfer to the university where you will pursue an engineering program, the most important thing to do first is to visit the personnel at the university who handle transfer credit. The pre-engineering personnel at the university can save you a lot of time and, therefore, money. There are, undoubtedly, many courses at the junior college that will not transfer for credit to an engineering program. If you take the proper courses, you can typically transfer up to four full-time academic semesters or six full-time academic quarters of credit, if the courses are taken at a regionally accredited institution. Little or no transfer credit is given for courses that are normally regarded as junior- or senior-level university courses and taken in the junior or senior years at the university.

The kinds of courses for which transfer credit is given are mathematics, chemistry, physics, English, history, humanistic/social science or equivalent-type university courses in this area, as well as some entry-level engineering courses such as statics, dynamics, strength of materials, and computer languages. All of these courses must be capable of satisfying the ABET criteria and/or university core course requirements. There are a number of courses that might appear to be eligible for transfer credit, but in reality are not. For example, any course with a title similar to that of one at the university but of a vocational nature will not be given transfer credit. A course taken at a junior college without a laboratory, that is similar to one at the university that contains a laboratory, will not be given transfer credit. Technical writing at a junior college is another course that is not eligible for transfer credit. At the university, courses in technical writing typically require as prerequisites both composition and junior or senior standing in class, because of the technical nature of the class papers.

Transfer credit is awarded for courses that satisfy the proper criteria, provided that the courses are equivalent to those taken at the university with respect to prerequisites, time, and content. Furthermore, a minimum grade of C or better must be achieved in each course.

If there is some question about the quality of a course you have already taken, you will probably be asked to bring in the book used in the course and the course outline. The faculty who would normally teach the equivalent course at the university will decide whether to award credit for the course you have taken.

Some course credit is awarded on an hour-for-hour basis, and some is awarded on a course-for-course basis. If there is a deficiency in the course hours to be transferred, for example, in transferring semester hours to quarter hours and vice versa, then additional hours may have to be taken to compensate for the deficiency. For example, suppose two three-hour semester courses taken at a junior college are to be substituted for two five-hour quarter courses at the university. Since two semesters is equivalent to three quarters, the two three-hour semester courses would transfer as nine quarter hours, and one additional hour might have to be taken at the university to satisfy the hour requirements in the engineering program being pursued.

Undecided Student

It is very important that students who are considering engineering as well as other curricula visit the pre-engineering personnel at the university. Suppose that you are considering engineering, architecture, and business. The university personnel can recommend to you courses that will transfer into almost any curricula, and therefore, you will not make the mistake of taking courses that will not count in any one of the programs of study under consideration. Taking classes at any institution consumes time and costs money, and therefore it pays to plan ahead to ensure that you are not wasting either.

Transient Student

This term defines a student at a university who wishes to take a very limited number of courses at other institutions. For example, suppose that a student wants to take one course at another institution near his home while he is home for the summer. Once again, the student should check with the university personnel first. The university personnel will typically want to know how the course will be applied to their program. If the course is a valid substitution for a required course, then permission can be granted to take the course, and a letter given to the student telling the hometown institution that the student is in good standing at the university. Three important items to re-member in these cases are (1) Prior permission from your university is required; (2) A grade of C must be achieved for transfer; and (3) Courses in the engineering major normally cannot be taken in this manner.

TIME TO COMPLETION

The university's bulletin or catalog will indicate that programs in engi-neering will take eight full-time semesters or twelve full-time quarters to complete, that is, four full-time academic (nine-month) years. While these statements are correct, they are often optimistic. Students who are intelli-gent, hardworking, and lucky can complete an engineering program in this time period. However, the average student normally takes a little more than four years. The reasons for this discrepancy are many and varied. For example, it is normally wise to begin an engineering curriculum by taking fewer hours than required in any semester or quarter in order to be sure that you are not in over your head. There may be scheduling conflicts with required courses, and therefore you may not get all the courses you need. There may be insufficient space available in the classes, and they may fill before you can get in. And, of course, you may not pass some classes and have to repeat them. Therefore, it is wise to plan ahead for a university ca-reer that takes longer than the minimum required period. You may even wish to extend your time in school. Taking longer than the minimum time to get a degree presents an opportunity to get a broader education that will benefit you throughout your entire career.

There is another issue that is worth mentioning here; that is, there are some five-year engineering programs. However, these programs normally include a cooperative education experience (to be discussed in some detail in Chapter 10), studies abroad, or some other added benefit that extends the time schedule. One such example of this added benefit is the growing trend toward a combined undergraduate and graduate five-year program that results in a combined Bachelor of Science/Master of Engineering degree.

5

Program Enhancement

ADVANCED STANDING/HONORS PROGRAM

Students entering the university with superior preparation or special competence in some specific area should check with the university prior to registration to determine their eligibility for advanced placement or credit. As an example, suppose that you have already taken differential calculus at your high school and that you have done very well in this subject. You may be eligible to skip the first calculus course, get credit for it, and enter the second calculus course immediately. This advanced placement may be granted based on any one or a combination of such things as the Advanced Placement examinations of the College Board, College Level Examination Program (CLEP) General and Subject examinations, departmental proficiency examinations, and the like. A word of caution is, however, in order here. Very few high schools offer courses in calculus, chemistry, and physics that are college-level equivalent. Therefore, if advanced placement is given as a result of an examination, there is a better chance of survival.

Somewhat aligned to advanced placement is the university's honors program. At universities where these programs exist, they are designed for gifted students capable of advanced work and provide a unique opportunity for academic excellence. Such a program is typically composed of a fixed number of students from throughout the entire university. Selection is based on high school grades and entrance test grades. In this program, special courses are taken in lieu of some of the required courses. In addition, Honors Certificates are awarded to the participants and the distinction of being a part of the program is also normally noted on a student's diploma and transcript.

ELECTIVE SELECTION

Most curricula have at least a small number of hours set aside for either free electives, technical electives, or both. This represents an opportunity for you to do some things that may be unique to your own interests and desires. Free electives are just that—free—you can take anything in the whole university for which you have the prerequisites. Thus, you have an opportunity to explore some areas that were not specified in your curriculum. For example, you can take courses in management, finance, language, art, music, and many other subjects that span the entire collection of subjects offered by the university. Remember that ABET specifies that you must take the equivalent of one-half year of humanistic/social science courses, and thus you may be able to complement these with your free electives if you so desire. However, you may also use these hours to pursue something that you have a flair for, such as guitar, piano, or the like.

Technical electives, on the other hand, are restricted to courses that support the major area of study. Even here, there is typically great flexibility. The courses in this category for engineering majors normally include engineering courses, both inside and outside the major, as well as courses in such areas as mathematics, chemistry, physics, and biological science.

For students who have a good idea of their technical interests, this is an excellent opportunity to broaden those interests or pursue some area in more depth. For example, a mechanical engineer interested in the computational aspects of machine design may take elective courses in such subjects as computer-aided design or computer architecture, which may be taught in computer science and engineering or electrical engineering. An electrical engineering student interested in cooling integrated circuits may take additional course work in thermodynamics, heat transfer, or material science, which may be taught in the mechanical engineering department. Keep in mind that junior college courses are not acceptable as technical electives because these are junior- and senior-level courses.

There is another aspect of elective courses that may be important to some students. When engineering students are within a very few hours of graduating (e.g., ten semester credit hours), they are typically permitted to take courses that can be counted toward graduate credit. Therefore, students who plan to pursue a graduate degree in the engineering major may take some courses that will become a part of their course requirement for an M.S. or Ph.D. degree. However, in order for these courses to count, you must be sure to have this plan approved by the faculty and/or graduate school in advance. Keep in mind that courses selected for use in a graduate program cannot be used for undergraduate credit also.

While the scenarios outlined above are certainly possible, in reality they are not always achievable. You may find that, due to such things as budget cuts and loss of faculty, the electives you want are not offered at all or not offered at a time when you can take them. However, if you plan ahead and check to see what electives will be offered when, you will be in the best position to get as much of what you want as you can.

ORAL AND WRITTEN COMMUNICATION

These are extremely important areas and ones that will definitely have a profound effect on your future. And yet it seems that few university students recognize just how valuable are these skills. Every now and then a student will request permission to avoid taking these courses in lieu of some highly technical subject. However, they are normally never permitted to make substitutions in this area.

Because both oral and written communications are so very important, you should accept every opportunity to speak before a group or write reports or other documents whenever you can, and if such opportunities don't arise, then search for them. This is one area where practice is very helpful. You may not be good in this area at first, but you will find that the more you practice the easier it becomes. You can be absolutely sure that if you will do as suggested, you will be delighted that you did.

The ability to communicate well is a real asset in college. Good oral communication skills will help you get the most out of every class, and good written communication skills will help you do the best in every class. However, the fundamental reason for feeling so strongly about the importance of these skills is based primarily on the critical role they will play in your future.

In many (if not most) cases, when you join the ranks of employed engineers, you will be assigned a task of writing a report describing your findings or solution. This single document will then be a mechanism by which you report your activities and accomplishments over some time period, which could be several months. If your superiors cannot understand your report, then one of the possibilities will be to question your viability for continued employment. After all, if you can't tell them what you have done, why did they pay you to do it? The ability to communicate effectively will be important in all your daily activities, which include writing letters or memoranda to, or conversations with, subordinates, associates, or superiors.

In industry, it is not unusual for high-level managers or directors to wait patiently for the first report prepared by the new engineers under their direction. Some young engineers have been in tears when their report was returned all marked up as if it had been graded. The time to learn to write well is when you are in school. The consequences of not writing well are less in a more benign college environment.

One of the most critical characteristics possessed by many captains of industry is the ability to communicate very effectively in a convincing manner, especially when speaking. They are able to talk to any audience and have the power to convince and persuade them. Furthermore, every superior leader in any field of endeavor has been an outstanding communicator. Whether you wish to become a leader in engineering or not, this is one of the most important characteristics that anyone can have, and the best time to hone these skills is while you are in school.

TEXTBOOK RETENTION

Some students have to sell their books used in one term in order to purchase the books for the next. If this is an absolute necessity, then so be it.

In addition, some students who could actually afford to keep their books sell them, because it is a source of funds that is essentially unknown to their parents. However, those students who can afford to keep their books should do so. The reason for this is straightforward. In many curricula in the university, the totality of the courses taken deals with a collection of areas that are not highly interdependent. For example, some liberal arts curricula consist of a large number of diverse courses that, in total, are designed to provide the student with a very broad-based education. In contrast, while providing breadth through the university core curriculum and ABET constraints, engineering curricula provide considerable depth, as illustrated, for example, by the typical semester-based (as opposed to quarter-based) electrical engineering curriculum shown earlier in Figure 4.3.

This figure indicates that the engineering courses build upon the mathematics, physics, and chemistry courses taken in the freshman year. The prerequisites, which will exist for the advanced courses, also provide an indication of depth. For example, the first few mathematics courses are stacked, and the student must pass Math n before taking Math $n + 1$. Furthermore, most of the mathematics courses are prerequisites, along with the circuits courses for the course in linear systems, which, in turn, is a prerequisite for courses in communications and control. Circuits is also a prerequisite for electronics. All courses in a particular area, for example, electronics, are prerequisites for the senior design courses—courses in which students are required to work in teams and apply much of what they have learned in different subjects to a complete design problem. Clearly, having a solid foundation upon which to build this multi-tier network is extremely important. You will find that quite often throughout your college career, as well as in your job after you leave the university, you will need to refer back to subject areas you have studied in the past. Although the university has a library that is well equipped with reference materials and your company may have one also, you are most familiar with the books you have used in your courses.

PERSONAL COMPUTATIONAL FACILITIES

Engineering students spend an enormous amount of time calculating and computing. Most homework, quizzes, and exams involve solving physical problems formulated as mathematical equations. Furthermore, as you progress through an engineering curriculum, these problems become progressively more complicated. Therefore, the more computational horsepower you have at your disposal, the better. A good scientific calculator is a must. You will need your own for tests, if nothing else. This technology changes rapidly, so check carefully before you buy one. Your professors can advise you on this purchase if you are in doubt about it.

There is one issue concerning calculators that should be mentioned. With today's modern alphanumeric calculators, there is sufficient memory to store all sorts of formulas, constants, and other relevant information that may be helpful on quizzes and final exams. However, some of your instructors may insist that you have this information memorized. If so, storing this information in your calculator for illegal use is a form of cheating. Therefore, it is important that you check with the faculty to find out exactly how they feel about the use of sophisticated calculators in class.

Universities have numerous rooms throughout the campus that contain either personal computers, engineering workstations, or both. However, if you can afford your own, buy one. In fact, there are some universities throughout the country where you have to buy one—it is part of the requirements for entering engineering. The software that is available today (some of it is actually free, and a lot of it is inexpensive) is capable of solving almost anything. Furthermore, the software scientists and engineers are working night and day to develop programs that will do everything except make the computer get up and walk around the room. Computers are here to stay, and the more familiar you are with them, the more help they can be to you.

A word of caution is in order here. Do not think for even a nanosecond (one-billionth of a second) that the computer is any panacea. It is essentially millions of very faithful devices that operate in a very methodical and precise manner under control of the software, using the data you give it. The speed of operation is unbelievably fast. However, it doesn't do any thinking. You have to do all the thinking and let it do the computation. In other words, the computer is anything but some all-knowing guru. This is an extremely important point, since many people take the attitude that because the answer was produced by the computer, it must be correct. The computer is simply an extremely fast calculator that computes exactly what you give it. Therefore, you must not lose sight of the importance of having some ballpark idea of what the results of a calculation should be, so that you have some check on whether your computer input and assumptions were reasonable. Failure by some to comprehend this point is the reason that decades ago when computers were introduced, people began using the acronym GIGO, which stands for garbage-in-garbage-out.

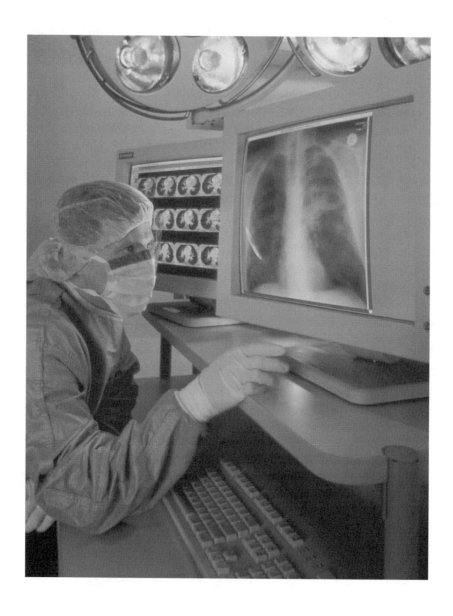

6

Involvement
in University Activities

HONOR SOCIETIES

Each curriculum in engineering has an associated honor society. For example, the honor society that recognizes outstanding students in chemical engineering is Omega Chi Epsilon, the honor society for civil engineers is Chi Epsilon, for electrical engineers it is Eta Kappa Nu, and for mechanical engineers it is Pi Tau Sigma. Other branches of engineering have similar societies. In addition, Tau Beta Pi is a national honor society that recognizes academic achievement in all areas of engineering.

Becoming a member of one of these societies is a mark of academic achievement and therefore a worthwhile goal. Each society has its own entrance requirements; however, they are all typically related to grade point average. If invited to join one of them, it is important to do so. Other reasons for this importance will be addressed later, but for now consider it a mark of distinction.

In addition to the honor societies that are directly related to specific engineering curricula, there are a number of honor societies that engineering students are invited to join because of their scholastic prowess, leadership ability, or service to the university community. For example, Phi Eta Sigma is a national honor society that recognizes high scholastic achievement for freshmen; Phi Kappa Phi is a national honor society that recognizes academic excellence in all disciplines; Mortar Board is a national honor society for outstanding seniors who have demonstrated distinguished ability in scholarship, leadership, and service; and Omicron Delta Kappa is a national honor society for individuals who have demonstrated exemplary character and specified eligibility in a specific area of campus life. This short list is but a sample of the honor societies that are active on most university campuses.

PROFESSIONAL SOCIETIES

All the programs in engineering have associated with them a professional society. For example, the professional societies for chemical, civil, electrical, and mechanical are the American Institute of Chemical Engineers, the American Society of Civil Engineers, the Institute of Electrical and Electronics Engineers, Inc., and the American Society of Mechanical Engineers, respectively. These organizations are involved in many activities that support their profession. For example, these professional societies work to advance the theory and practice of their area of technology through such activities as conferences, publications, educational programs, and standards. They publish career guidance materials, and they are key players in the accreditation of engineering programs through their active participation in the Accreditation Board for Engineering and Technology. Furthermore, they work to enhance the quality of life through the application of technology and promote an understanding of the manner in which technology affects the public welfare.

As a future professional in some area of engineering, you should get involved in the society that is related to your curriculum. This organization will enhance your education. Generally, your participation in the student chapter of your professional society will provide the opportunity for you to develop professionally, to increase your knowledge of engineering practice, and make contacts and friendships that will benefit you in the long term.

The immediate benefits that an engineering student gains from participation in the student branch/chapter of their particular professional organization are such things as the opportunity to meet and talk to professional engineers in order to learn how to apply their studies, to tour industrial plants, to work with fellow students on a variety of student projects and contests, and to serve in some official capacity in the organization to improve leadership qualities and communication skills. However, discussions with many students indicate that what is perhaps their primary motivation for joining a professional organization is the help that this organization can provide in obtaining employment—both summer jobs and career positions. This is normally accomplished through professional contacts and such activities as organization-sponsored job fairs. Finally, it is important to note that industry is interested in people who are interested in their profession, and membership and participation in a student professional organization is an excellent way to demonstrate that interest.

STUDENT ORGANIZATIONS

There are a wide variety of student organizations that exist on most campuses, and they will normally include one or more of the following: student government, service organizations, special interest groups, sports clubs, and religious organizations. The student government is the body that controls all student projects within the university. These types of organizations are typically modeled after the U.S. government and therefore contain executive, judicial, and legislative branches.

This organization is normally funded through student fees, and therefore all aspects of the service performed are open to all students. The organization is a powerful lobbying force for students and is the driving force behind a host of projects, including campus publications, recreational services, entertainment, special events, and a variety of other activities, such as blood drives for the Red Cross, and the like.

Most campuses have many service organizations that welcome student volunteers. They typically organize and execute a number of projects that benefit the students as well as the campus community. One such organization is Circle K, which is sponsored by Kiwanis International.

Engineering students, like other students on campus, can become involved in a wide variety of extracurricular activities. Students may choose from a smorgasbord of organizations that deal with essentially any activity of interest. For example, some of the organizations are Amnesty International and Habitat for Humanity. Students can also work on the student newspaper, radio, or yearbook.

Universities typically have a large number of sports clubs that students can join in order to participate in a sport of interest to them. For example, some typical organizations are clubs for fencing, football, karate, lacrosse, racquetball, soccer, and tennis.

Most campuses have a large group of student organizations that are linked directly to religious denominations. They sponsor a variety of religious and social functions that foster the student's continued spiritual development.

FRATERNITY/SORORITY

Fraternities for men and sororities for women exist on most university campuses. They are primarily social organizations that are coordinated through three groups. The governing body for fraternities is the Interfraternity Council, for sororities it is the Panhellenic Council, and for the traditionally Black fraternities and sororities it is the National Panhellenic Council. The organizations set their own selection criteria, and most students that join these groups do so in their freshman year. In fact, some organizations have quotas for students beyond the freshman year, and therefore students interested in joining one of these organizations are encouraged to do so early.

The sorority selection process, organized by the Panhellenic Council, is a structured formal rush week. Anyone can go out for rush, and the actual selection is a mutually agreed upon decision. In other words, the sororities select students, and the students select sororities. The students can then select the one that is, in their opinion, the perfect match for them. The rush sponsored by the Interfraternity Council and the National Panhellenic Council is typically less formal and not highly structured. Otherwise the process is essentially the same. If you are interested in joining a fraternity or sorority, be prepared to ask a number of questions, such as: do they stress academics, and where are they ranked at the university with regard to this criteria. Another important question is cost.

Many of these organizations have houses and, in some cases, this can be the cheapest way to live. These organizations provide a home away from home. It can be a support system where you can build bonds with other students of similar interests.

TIME CONSIDERATIONS

An item that was not addressed in describing the organizations in this chapter was time. Unfortunately, there are only 24 hours in a day, and engineering students quickly find that their studies take up most of that time.

Therefore, students must prioritize their time and decide how much of it they wish to devote to various activities.

Becoming involved in a variety of activities is certainly recommended. It is a very broadening experience, which permits interaction with a large segment of the university student population. It also helps to make you a more well-rounded individual, and it tends to prevent burnout, which can result from a tunnel-vision approach to your course and laboratory work. Furthermore, more than one study has concluded that student achievers in extracurricular activities are the ones who best succeed in life beyond the university, assuming that they have acceptable grades in their courses. However, the more extracurricular activities in which a student becomes involved, the fewer academic hours that should be attempted per academic term.

Finally, if a lack of time becomes so serious that it begins affecting your health, it is important to note that all campuses that are accredited by their regional accrediting bodies (e.g., the Southern Association of Colleges and Schools for the Southeastern United States) must have a variety of services and programs that will assist you in coping with problems of all types. Some students experience personal crises, emotional stress, or just have difficulty making the transition to college life. A personal assessment, counseling, or health center is available to help with all these issues. Your advisor is also someone who can identify for you various sources of help.

7

Understanding the Faculty

FACULTY ACTIVITIES

The faculty is the university! Although this statement may surprise you, it is absolutely true. The administration, support personnel, and facilities such as buildings are present only to help the faculty do its job. The university therefore develops its reputation primarily from the faculty and to a lesser extent from its graduates. The faculty in turn develops its reputation from the scholarship (i.e., research and publications in national and international journals) as well as their leadership in professional societies and participation in professional activities, such as technical conferences. The truth of this statement helps explain a number of things about a university.

The faculty members who serve at state-supported universities, and to a lesser extent in other types of institutions, have numerous responsibilities. In general, they must work on three fronts—teaching, research, and public service. Some institutions may not be responsible for the third element, but most of them will definitely be involved in both teaching and research.

Teaching involves preparation for classes as well as maintaining office hours when students can receive individual help with questions, problems, or some portion of the course that is difficult for them to understand. In preparing for their classes, they must assemble lecture notes, handouts, view graphs, slides, and anything that will enhance instruction, for example, a laboratory demonstration as shown in Figure 7.1. Quizzes and homework assignments must be prepared, assigned, and graded. In addition, some of the instruction may be provided at off-campus sites in other cities.

The faculty members' second responsibility is that of performing research that advances the frontiers of knowledge in their area of expertise. First of all, the faculty members must find the money to support themselves and the students, normally graduate students, who work with them. They must be entrepreneurs and hustle the funds. You can be absolutely sure that no one is standing just outside the university with a sack of money begging the faculty to take it. The faculty must prepare proposals and compete not only with faculties in other universities, but often with in-

Figure 7.1. Professor discussing the operation of an electric power system with a student.

dustry, too. Faculty members who are good at this exercise not only enhance their own reputation, but that of the university as well, since this is one of the indices used to rate universities.

You may find it interesting to learn that some, if not most, of the universities in the country that are state-supported only receive a portion of their operating budget from the state. For example, typical funding for a state-supported institution might be one-third from the state, one-third from tuition, and the remaining one-third from research. Note carefully that your tuition only pays a fraction of the cost of your education! In this environment, the professors must find sources of money to support many activities in the university. The research actually supports the teaching program. For example, what is learned in research often finds its way into the undergraduate program to keep it up to date. For instance, the electronics faculty will ensure that its portion of the undergraduate program teaches the modern integrated-circuit electronics and not out-of-date technologies such as vacuum tubes.

The third responsibility is that of public service. The faculty is expected to support in every way that it can the development of the state's citizens and, in particular for engineering faculty, its industry. This interaction may take a number of different forms. For example, the professors may work directly with industry on some project, or they may provide short courses—typically a few days in length—to industry personnel in an attempt to keep them up to date.

In addition to the responsibilities listed above, the faculty members are typically involved in a variety of other activities. Some work with student groups (e.g., the student branch of the professional society) in the development of meetings, field trips to industry, or projects. Some work with honor societies, or professional societies on a regional, national, or international basis. For example, a mechanical engineering professor may be involved in the activities of the Accreditation Board for Engineering and Technology (ABET), the American Society for Engineering Education (ASEE), the American Society of Mechanical Engineers (ASME), and the National Society of Professional Engineers (NSPE). This involvement is not only good for the professor's career but, in addition, the university's name is carried with him or her. If the faculty at a university is well known, then there is a direct benefit to the students, that is, if your prospective employer is familiar with the professional reputation of your professors, that reputation will carry over to you, to some extent.

Finally, faculty members are like every other individual in the community. As such, they are involved in other educational activities in K through

12, as well as civic and religious organizations. In general, they are most supportive of the students. After all, they have chosen a profession of service, and they are ready to help you in any way they can, especially if you ask.

GRADUATE STUDENT TEACHERS

Some students find that a significant number of their classes are taught by graduate teaching assistants. Graduate teaching assistants (GTAs) are typically used as graders and often teach the laboratories, but normally only a select few are permitted to teach classes. In general, the GTAs that teach lecture courses have to be essentially Ph.D. candidates, since the accrediting agency for the university typically has this requirement as a criteria.

Graduate students are employed to teach in a variety of capacities, primarily due to a lack of funding. If the departments could afford it, they would have faculty members doing everything. However, life just isn't that simple. By employing GTAs instead of faculty members in some courses, the cost of education is kept down, and equally important, the GTA is helped both financially and intellectually.

Some students and parents fear that classes taught by GTAs do not provide as good a learning experience as could be obtained from a professor in the classroom. There are some cases where this is certainly true. However, it is not true in general. The GTAs that teach are usually screened and are prepared to do an excellent job in the classroom. In fact, you might be surprised to learn that, in many cases, the students rate some of the GTAs as better teachers than some of the professors.

An important point to keep in mind is this: If you have a teacher that is really ineffective, there are normally two things that you can do. First, visit the head of the department in which the course is taught. After all, this individual may not know that your teacher is doing a poor job. In addition, most universities have teaching evaluation forms (see the next section), and these can be used anonymously and effectively to inform not only the teacher, but his or her superiors, that a problem exists.

GRADING THE TEACHERS

The grapevine in the university is short and fast. It doesn't take long to learn what some students think of their teachers. An incident that will not soon be forgotten occurred to some freshmen in an English class. On the first day of class, the students sat patiently waiting for the teacher to arrive. When she did, one of the students in the first row got up, stood in front of the class, and announced "Oh no, not me, I'm not staying in here." His comments turned out to be sage advice. Those who remained in the class wished that they had not. However, there is always plenty of advice floating around, and if you receive some of it, if possible, try to calibrate the individual who is supplying you with these words of wisdom. If this student is a party animal, has a poor GPA, and rarely if ever studies, the entire university faculty may be on his or her hit list. Furthermore, there will be situations in the university, just as there are throughout life, that will be difficult. You cannot simply run from them. Running has to be the absolute last resort. When your life is viewed in toto, the university is a rather benign environment, and therefore it is a good place to start learning how to make the best of difficult situations.

As indicated earlier, you will normally have an opportunity to grade your teachers through the use of a university-wide teaching evaluation form. These forms are handled differently at various universities. For example, in some cases, the results go to the teacher, the department head, and perhaps the dean of the college. In some universities, the results are published by the students for general distribution. Regardless of the exact manner in which they are used, they are usually an effective feedback mechanism.

Like anything else, if they are used properly, the teaching evaluation forms can be useful tools. Unfortunately, sometimes they are not. Some students use it as an opportunity to try to harm a teacher they dislike for some reason. In this case, it becomes a retaliatory mechanism for a bad grade, which may be well deserved. Furthermore, remember that the teachers were not born yesterday, either, and they can gear the course to the teaching evaluation, which may not provide you with the best learning experience.

Always keep in mind that the real mark of a good teacher is how much you have learned, not your grade or the difficulty of the course. Many students, long after graduation, refer to some of their hardest teachers with reverence because that teacher had a real impact on their job performance. And, after all, that's what it is all about. You attend the university to get a good education so that you can lead a happy and productive life.

8

Personal Performance Enhancement

GOALS

If you decide to leave the university but do not care where you are going, then any road that exits the university will do. However, if you have some specific destination in mind, there may be a very small number of roads that will take you there, and there is probably one road that is better than all the others. In addition, if your destination is far away, it would be ideal if you were enthusiastic about the trip and had all the means necessary to get there. If you think about it for a moment, achieving your goals is very similar to making such a trip.

Your goals, while you are attending the university, may span a range of activities. For example, your goal may be to make an A in mathematics, a B in chemistry, and an A in English composition during your first term in school, achieve a specific grade point average (GPA) by the end of your freshman year, qualify to join the Tau Beta Pi National Engineering Honor Society, or maintain a specific GPA while serving as vice president of the student government association. Such goals as shooting in the low 70s in golf or some similar activity have been deliberately excluded because you won't have ample time for these types of activities.

Regardless of the type or number of goals you have, they should be simple to define, very specific, and measurable. Your goals should not be nebulous; if asked, you should have no difficulty explaining or defining them. For example, if you told us "you wanted to be a better student than you are now," one would be prompted to ask "what does that mean?" How will you ever know if you have achieved it?

Your goals should also be realistic and achievable. They need not be easy or straightforward, but there must be some way, no matter how complicated or difficult, for you to achieve them. If you have selected a goal that simply cannot be achieved by any means, then you are doomed to failure before you start. Therefore, a realistic assessment, up front, of what is actually involved in reaching a goal is extremely important. A naive approach could easily result in a waste of time and resources. The naiveté with which some people approach goals is illustrated by a conversation with astronaut Hank Hartsfield in which he commented that an individual with whom he was speaking could not understand why going to the moon was such a big deal—after all, you could see it all the way!

Planning is a key factor in both setting and attaining goals. In fact, planning is an absolutely necessary precursor for accomplishing almost anything. That is why, for example, contractors have completion targets, factories have production schedules, and salespeople have quotas. The achievement of any worthwhile goal normally takes time, and hence you must develop a plan and execute it in order to be successful. Having a crystal-clear image of exactly what you want to accomplish and a well-defined plan for achieving it are equivalent to firing a heat-seeking missile at a "hot" target. Your internal guidance system will automatically provide in-flight corrections all along the way. This homing device continuously repositions you for an optimum path to the goal, even in the event of sudden setbacks or diversions that are encountered along the way. It is impossible to overemphasize the importance of having a clear "fix" on your goal, for without it you will simply drift aimlessly, and the probability that you will somehow stumble into it is too remote to warrant even the slightest consideration.

You should look at your overall plan as a sequence of events involving sequential decision points. As a freshman, you should plan to successfully complete the freshman engineering or pre-engineering (or what is also commonly called the preprofessional) program, which includes course completion and credit hours earned for degree credit. Successful completion of this work leads to a decision point. At this point, you have partially completed the university's core requirement, determined your comfort level in these courses (i.e., "I can really do this stuff well"), and thus can decide to go on to a specific branch of engineering, or enter some other curriculum. While the work is in progress, you should ask yourself if you enjoy what you are doing. The answer to that question should guide you at a decision point that comes no later than the end of the freshman year. If you enjoy your classwork and are doing fine, then you have several viable options. However, if you are not comfortable with the pre-engineering work, how can you be comfortable in advanced courses that build on this broad base of course work provided in the freshman year? The goal is not to get by some particular portion of the coursework, such as mathematics, but to build a solid foundation upon which to launch a successful engineering career. Some students know exactly where they are going, but do not do well in the freshman course work. These individuals should examine their situation to see if they have the wrong goals.

As you pursue your goals as a student in engineering, the skills you develop will transfer to those situations where you are required to enhance your professional development by continuing to learn. Your multifaceted professional career may even include leaving engineering to do something entirely different. However, you will have the tools, such as discipline and the ability to think critically, to make a successful transition.

TIME MANAGEMENT

Other than your God-given talents, time is the crucial variable that you have to allocate to accomplish your desires and goals. Each person has the same 86,400 seconds in a day to spend as he chooses. Your success, in school as well as in your career, is critically dependent on what you do with those seconds. If you budget your time wisely and do not waste it, you will not only have time to do the things you need to do, but will have time to do the other things that enhance your life.

While in high school, a student will normally spend 35 hours per week in class and approximately 10 hours per week outside of class studying and doing homework, for a total of 45 hours. Assuming a class load at the university of 15 hours, experience indicates that if 30 hours per week are spent in dedicated study, for—once again—a total of 45 hours per week, the student should be able to pass. Note that the operative phrase here is "dedicated study." Some students will sit down to study for a couple of hours and after a few minutes of work will get up to make a sandwich. After a few more minutes of study they decide to work out with weights. Two additional short periods of study interrupted by telephone calls and a favorite TV program complete the study period. Unfortunately, it is not unusual for students in this mode to claim they have been studying for hours!

Budgeting time implies some effort in working to make things happen, and thus is not a passive activity. Students who approach their educational programs in a passive, rather than an active, mode are like the persons who are "waiting for their ship to come in." Notice that the difference here is "waiting," rather than "working." These students will never see their ship in the harbor, simply because there are just too many people who are "working" to get their ships in, and there is no way without some effort that these students will ever make it.

In time management, organization is the critical factor. Through careful planning and scheduling, you place yourself in better control. By scheduling events in a realistic fashion, you are able to manage your time as efficiently as possible, even in the face of uncertainties. Organized management of your time means setting priorities, avoiding procrastination, and eliminating confusion. The following suggestions are helpful in this regard.

Schedule your affairs so that you can work problems for as long as possible within the required time constraints, and you will find that your subconscious will help you solve them. For example, if you are assigned a term paper that is due in several weeks, start on it immediately, if only to plan the attack. If unforeseen problems surface along the way, you will have time to recover from them. However, if you wait until the last minute to start, there may be no time to adjust. Keep a set of 3" × 5" cards or some similar mechanism with you at all times so that, as ideas pop into your head, you can write them down before you forget them. Order the tasks you have to do by priority, and do the most important ones first—and do them correctly. After all, if you don't have time to do it right the first time, where are you going to find the time to do it over? You may even need to reorder the task list more than once a day. Reordering the tasks in real time permits you to optimize the schedule either because of additional data or new—more important—problems that have just surfaced. You may even be able to sandwich small jobs in between big ones in order to optimize your schedule. For example, if you have time between classes, don't waste it; do something you have to do that only takes a small amount of time.

Some students try to study all night for a test. This is almost always a mistake. It is a misuse of time because, after a while, everyone reaches the law of diminishing returns. In engineering, it is most important that you are able to think, and you can't do that if you are exhausted. In fact, there have even been cases where a student who stayed up all night studying came to the exam and promptly fell asleep.

You have to be smart in dealing with your fellow students so that, while maintaining a congenial environment, you do not permit your colleagues to waste your time, since you only have a fixed amount of it. One of the executives at NASA in Huntsville, Alabama, had an unusual way of handling people who dropped by to provide an unsolicited memory dump. Under the carpet next to his desk he installed a switch that activated a buzzer at his secretary's desk in the next room. When he decided that some conversation had gone on long enough, he would, unknown to his visitor, press the switch with his foot. His secretary would pop into the room and inform him that it was time for his next appointment.

Finally, you should ensure that interspersed within your work time is time for some rest and relaxation. These periods will clear and refresh your mind. For example, it is not unusual for a student to work for hours on some problem without arriving at a solution because of some error in the method. However, after taking a break, for example, going to a movie, playing tennis, or whatever, she returns to the problem and immediately recognizes the error. What seemed to be invisible earlier is later obvious. Everyone invariably reaches the point where the law of diminishing returns takes over, and when this happens, you are actually wasting time if you continue to press on. In fact, one technique that seems to be an effective incentive in the optimization of your time is to reward yourself with some pleasurable activity as soon as you have completed some difficult or unenjoyable task.

Time is one of your most precious commodities and, as such, you should treat it with a great deal of respect.

STRESS MANAGEMENT

Stress is typically produced by either overload, conflict, or uncontrollable conditions. You may try to do too much in too little time, deadlines may be too short, or standards may be too high. You may try to be someone that you are not, or you may simply be the victim of circumstances that are completely beyond your control.

Students in a university encounter a wide variety of situations that can cause stress. For example, they may sign up for too many classes and be unable to handle the load, tests in all their classes may occur on the same day, they may be involved in too many extracurricular activities and thus not have the time to keep up with their studies, there may be girlfriend/boyfriend problems, they may fail to keep their finances in order, and lots of other problems, some of which may be thrust upon them with little or no notice. Occasionally, parental pressure leads students into a particular curriculum, one in which they are not really interested. Their parents probably had only the best of intentions; however, their parents did not have to spend at least four years preparing for a career that they did not want. Such a situation can be very stressful and normally ends in disaster. It is unfortunate, but stress will not end with graduation. It is with each individual throughout their career, and one must learn to control it and deal with it.

Stress can have either a positive or negative impact on you. If you are unable to manage stress, your reaction is normally worry, anxiety, or depression. In fact, some students who find themselves in a very stressful position drop out of the university and, many of them never return. However, well-managed stress can be a very effective mechanism for growth, which in turn permits you to handle even higher stress levels. Reasonable levels of stress, when kept under control, can actually maximize achievement.

Your attitude is perhaps the single most important factor in your ability to conquer stress. As Milton so eloquently stated in his poem, *Paradise Lost*, "The mind is its own place, its own heaven, its own hell." Regardless of the actual circumstances, if you think something is bad, it is bad; and if you think it's good, it is good. Hence, you should always try to look at the bright side of everything. While it is impossible for you to control all the events of your life, the one thing you can control is what you think about them. It is your perception of the impact of events on your life that is really the crucial issue.

Finally, as a student at the university, you will be forced to accept things over which you have no control. For example, classes may not be available when you want to take them, or the university may raise the tuition. You must learn to accept the things over which you have no control. Since you can't do anything to change the situation, any energy you expend in worrying about it simply goes up in heat and is detrimental to your well-being. When this type of situation arises, you will be forced to accept it and go on. As a child, you probably read the following Mother Goose rhyme, which applies so well in this context.

> For every ailment under the sun,
> there is a remedy, or there is none.
> If there be one, try to find it.
> If there be none, never mind it.

NETWORKING

In the present context, what is meant by networking is the interconnection of friends, acquaintances, or anyone else that can provide you with support in some aspect of your career. If you have not tried it, it will literally amaze you how quickly you can obtain some unknown information through a network of people. As a simple example of this concept, consider the young man in high school who has seen a young woman from another high school across town that he would like to date. In addition, he has heard that this young woman is very strict and will not date someone either she does not know or who has not been recommended by someone she trusts. However, if he is able to find one of his own classmates that knows him well and who may, in turn, have a friend at the other high school who knows the young woman in question well enough to provide the critical recommendation, then a network can be employed to establish the desired connection. People are interconnected in a wide variety of ways. In fact, there is often a great deal of overlap in this interconnection network. Some of the possible interconnections occur through family ties, individuals who go to the same church, work in the same building, belong to the same civic organizations, are members of the same professional organizations, graduated from the same school, and on and on. Such a network can provide you with a host of information, including sources of information for your studies, job possibilities, advancement opportunities, and the like. The potential of such a network is absolutely amazing and it is wise to start early developing your own, since it will become even more important after graduation.

9

Paying for Your Education

COOPERATIVE EDUCATION PROGRAM

What is Cooperative Education?

Co-op education was started in the United States in 1906, and today there are perhaps thousands of students in co-op programs at educational institutions throughout the nation. The program is a triad relationship among the student, the university, and an employer, in which engineering practice in a work environment is combined with academic study. The basic approach is to alternate academic terms between schoolwork and professional work. The co-op work is career-oriented on a professional level. Employers involved in the program provide progressively challenging assignments as an individual proceeds through the program. Thus the co-op work augments the student's education by introducing practical experience. Most universities have a co-op office that handles all the details of the program and helps qualified students select an employer suitable to their needs and desires.

Advantages of the Program

Although the co-op program typically extends graduation by at least one year, there are numerous advantages, both tangible and intangible, to individuals who participate in the program. For example, employers can normally provide educational resources not found at the university, resulting in a more valuable educational experience. Figures 9.1 to 9.5 provide an indication of some typical work performed in a co-op assignment. Thus your educational experience is not only enhanced through the acquisition of new specialized knowledge, but you have the opportunity to see it applied in a real job situation. This added dimension to your education stimulates your learning while in school and is thought to account for the improved attitude that typically exists among co-op students. In addition, the time spent with an employer helps you better define your career goals, builds your confidence in your ability to function in a working environ-

Figure 9.1. A co-op student testing propulsion modifications to the General Dynamics (now Lockheed Martin) F-16 Fighter Aircraft. (Courtesy of NASA Langley Research Center)

Figure 9.2. A co-op student surveying for a new manufacturing plant. (Courtesy of BEC/allwaste)

ment, teaches you some very valuable lessons in the human relations aspects of the workplace, makes the transition to work at graduation a smooth one, and may permit you to achieve a higher starting salary at graduation as a result of your work experience. Finally, although the primary concern of a co-op program is educational and not financial, the salary you obtain while working will help defray many of the expenses at school. Obviously, the amount of money saved during the work terms will be a function of many factors, not the least of which is your ability to discipline yourself and operate in a thrifty mode.

The Triad Relationship

A co-op program is a very close relationship among the employer, the student, and the university, and each must do its part in order to ensure a successful program. Whether the employer is a government or an industry,

Figure 9.3. Engineer and co-op student discussing substation equipment.

it is responsible for providing you with a work experience in your field of interest that gradually becomes more challenging as you progress in the program. You will probably work under a professional who will direct your efforts and assign you new tasks that support your professional growth. Depending on the employer and your own field of interest, the assignments may run the gamut from research and development to manufacturing. In your final work term, you should be working at the level of a junior engineer.

Employers who participate in co-op programs treat their involvement as a long-term investment. They are able to involve bright, young, and energetic university students in their operations. During the time that you are working with them, they are able to evaluate your potential for permanent employment with them. However, participation in the program does not obligate either you or the employer to a permanent relationship.

Figure 9.4. Engineer and co-op student monitoring a centrifugal machine in a chilled water plant.

As a co-op student, you will be working in a professional environment and as such you are expected to adopt professional standards of conduct. You should do the very best that you can as a dedicated and responsible employee. As a partner in the triad, you are able to play a significant role in the selection of your employer and, therefore, you are obligated to remain with that employer throughout the co-op program. In addition, you must fulfill the normal requirements of an individual in a co-op assignment by preparing the required reports as well as maintaining your grades to ensure continuous eligibility.

The university personnel who manage the co-op program are responsible for obtaining employers who will provide you with a co-experience that enhances your educational objectives and arrange for interviews with these prospective employers. They also counsel you throughout the program and ensure that you are always in compliance with all university policies.

Figure 9.5. Co-op student employing a computer-aided design (CAD) system to design a microelectronic circuit.

Qualifying for the Program

If you believe that co-op education is viable from your point of view and you wish to pursue the program, you may do so upon being accepted for admission to the university as either a freshman or transfer student. Once you have demonstrated your academic ability during at least one term at the university by achieving at least a C+ average, you can be placed with a mutually agreeable employer. In general, you should be at least 18 years of age, a U.S. citizen, and physically fit. Although the program does not disallow married students, it is more difficult for these individuals because of the frequent changes in residence. Students with military obligations also find the co-op program difficult, if not absolutely impossible, because of time-constraint conflicts. Finally, in addition to requiring a C+ average for entering, this grade point average requirement is also necessary for continued participation.

The Co-op as a Student

Co-op students are considered to be full-time students, even while on a work term, and therefore eligible to participate in all extracurricular university activities, such as sporting events. Because the program is an integral part of the educational process, the work experience gained in the program becomes a part of your transcript and, as such, a part of your permanent university record. In addition, the transition from school to work seems to be easier for those students who participate in the program. Finally, as the nation pushes toward greater diversity in the workforce, the opportunities in this program for women and minorities are numerous.

PART-TIME EMPLOYMENT

If you are aggressive, there are typically a number of jobs available both inside and outside the university that can be scheduled around your classes. For example, the university normally has what is called Work-Study funds that are supplied by the federal government to support students who qualify for them. However, the university will also have its own funds for supporting students who are willing to work in a variety of capacities. Some potential jobs that are available to students include running errands and operating as a technician aide within a department, or performing maintenance of all types for the university as a whole.

The local economy may have many temporary employment opportunities that are available to students. These jobs may range from one end of the work spectrum to the other. For example: general maintenance, including cleaning, painting, and landscaping; retail or department stores; grocery stores; automobile service stations; and fast food restaurants are just some of the places where students can work to earn some of the money needed to support their education. Some students have talents that can be explored in these situations. For example, an individual who can sing or play a musical instrument may be able to join a group that performs locally.

During the January-to-March time frame, numerous employers advertise summer jobs or internships through the Career Counseling and Placement Center, an organization that will be discussed in some detail later. In addition, recruiters who are on campus to interview students for career employment may also interview students for summer jobs. These summer positions may be positions in engineering firms or they may be jobs at camps, resorts, theme parks, and the like that are not related to the engineering major. There is something to be gained from any of these positions, in addition to the money they provide.

RESERVE OFFICERS' TRAINING CORPS (ROTC)

Students interested in a military career could potentially be selected for a ROTC program. The ROTC requirements must be integrated into the curriculum requirements and may increase the length of time in college.

The military services have recognized the fact that modern weapon systems are becoming very sophisticated, high-technology systems. Therefore, they are normally very interested in attracting more engineers to their ranks, and therefore engineering students are sometimes given preferential consideration in the screening process.

All branches of ROTC (i.e., Air Force, Army, and Navy) provide a mechanism for helping you get through school. While the Air Force seems to be more interested in electrical and computer engineers, the other branches are interested in all disciplines of engineering. Of course, you must pass both a mental and physical test to join; however, if selected for a program, you are typically paid for your tuition and books, plus a monthly stipend to help cover other expenses. Some private universities even supplement room and board. Because engineers normally do not graduate in exactly four academic years, the services have the flexibility to extend your benefits for a longer time. However, these extended benefits are not automatic and are given only to those cadets who are making excellent progress in their studies.

When you complete an ROTC program, you are commissioned as a Second Lieutenant. Your service obligation is four years. The position to which you are assigned will depend on the needs of the military but, to the extent possible, you will be assigned to a position that matches your background and interests.

SCHOLARSHIPS

The university admission forms normally have a space on the application where one can indicate an interest in scholarships and financial aid. Scholarships are normally based on merit and/or need. Some universities are capable of awarding scholarships up front based solely on high ACT or SAT scores and a high GPA. Individuals with very high ACT or SAT scores and excellent grades will typically be contacted by the universities. For example, national merit scholars receive $750 from the National Merit Corporation and then the universities begin bidding for them. The approach goes something like this: if the student will list a particular university as the first choice with National Merit Corporation, then that university will give the student a grant to pay tuition, or tuition and room, or tuition and room and money toward books, or all of the above plus a quarter or semester of study abroad. This is obviously a lucrative business for the brightest and best students.

University scholarship applications are normally mailed out in the late fall or early winter. Typically, universities make their decisions on their best scholarships within a couple of months, while others may take longer in their deliberations. Once the first-tier scholarships have been awarded, there are usually a number of other scholarships that remain. These second-tier scholarships may be general scholarships, which are specifically designated for students in a particular curriculum, or they may be scholarships that are awarded by some alumni group and aimed at students that come from a specific community.

In addition to the general scholarships awarded by the financial aid office of the university, scholarships may be awarded by the dean of engineering's office, specifically for engineering students, or there may be departmental-designated scholarships for students who wish to major in a specific discipline, such as electrical engineering. Therefore, both the dean of engineering and the head of the department you wish to enter are also possible sources of support.

Students who are interested in going to school out-of-state should contact the university of their choice to see if there is a program for waiving out-of-state tuition. Some schools will waive the out-of-state fees for students with a reasonably high ACT (or an equivalent SAT score), and some universities will waive all or a portion of the out-of-state fees for the sons and daughters of alumni if they have an average ACT score.

Scholarships vary widely from university to university, and therefore it pays to look into this aspect very carefully.

One of the newest sources of funds has resulted from money collected from gambling in some states. A case in point is the state of Georgia, which uses its lottery money to support its residents to go to in-state schools. In these types of programs, students are typically required to maintain a certain level of academic performance (e.g., a B average); however, as long as they maintain the necessary grades, the state pays their tuition. Furthermore, there may be some limit on family income in order for students to qualify, but not necessarily. The program in Georgia is called HOPE—Helping Outstanding Pupils Educationally.

There are many other sources of scholarships in addition to those that are controlled by the university. The professional societies listed in Chapter 1 also give scholarships, and it pays to contact the ones in which you have an interest. Since 1985, the National Action Council for Minorities in Engineering (NACME) has provided approximately $40 million in financial aid for minority students in engineering. As a result, more than 4,700 minority students have graduated in engineering with NACME support.

STUDENT LOANS

Students who have demonstrated a sense of responsibility can often obtain loans that can be repaid after graduation. These loans may come directly from parents, relatives, or friends. They may also be obtained from some banks, if someone close to you is willing to co-sign the note. The U.S. Government may also provide some assistance. To obtain information on government assistance, you may contact the following office.

Federal Student Financial Aid Information Center
P.O. Box 84
Washington, DC 20044
(800) 433-3243

TUITION EXPENSES

The cost of an engineering education normally spans a range that extends from that needed as a resident to attend state-funded/-supported universities on the low end to that required for private universities on the high end. Although the numbers given in Table 9.1 are approximate figures, they do provide reasonable estimates. Obviously, there are many factors that affect some of these numbers, for example, is the university located downtown in a big city or in a small rural community. It is important to note that these figures are in 1995 dollars and thus will change over time. Some schools might increase their tuition every year while others will hold theirs constant for some time and then have an increase. Either way, it is probably reasonable to count on an average increase of approximately five percent per year.

As the table indicates, the cost of an engineering education will range from approximately $30,000 to over $80,000, at a minimum, depending on the particular university attended.

A word of caution is in order here. If the student brings a car to school, there may be a number of expenses that are not included in Table 9.1. Is the car paid for? If not, car payments can be a significant financial drain. Regardless of this item, there are registration fees, insurance, routine maintenance expenses, and oil and gas. Experience indicates that this item must not be overlooked and has actually been for some students the roadblock to the successful completion of an engineering program.

Table 9.1   Academic Year Costs

	Resident Attending a State-Funded/-Supported University	Attending a Private University
Fees	$3,000	$18,000
Room and Board	$4,000	$6,400
Books and Supplies	$650	$650
Personal Expenses	$2,000	$2,000

One final point to consider before leaving this chapter is the difference between in-state and out-of-state tuition. Remember the people in any given state pay taxes to support the public institutions in their state. Therefore, if you want to attend a university that is outside the state in which you live, you will pay a premium for that privilege. Tuition is typically several times that quoted for in-state students. This can be an important factor in considering which university to attend. Keep in mind that, as indicated earlier, there are situations in which this out-of-state fee is waived.

Private universities are a different situation altogether. Because there is no state support, students must pay the entire cost of education, and, therefore, the tuition in these institutions is always quite high.

RL10A-4

10

Career-Oriented Issues

POSITIVE MENTAL ATTITUDE

As you approach graduation you should do so with a positive mental attitude. You may not remember everything that you have learned and, in fact, most students don't. However, you have learned how to think critically, and you have a solid background in mathematics, science, and engineering. Most employers will not expect you to be in a position to produce results right away. In fact, many of them will put you through some sort of training program to acquaint you with their processes and products. Except in rare circumstances, your employer will have already hired numerous other engineering graduates and thus will have a keen appreciation of your level of knowledge when you graduate from the university.

It is important to remember that you are completing one of the most demanding curricula in the university and, therefore, are well prepared to launch an exciting and rewarding career. The attitude that you carry into this endeavor will profoundly affect your resultant success.

There is tremendous power in your attitude, and because of the enormous control that it exhibits over your every action, it is extremely important that your attitude be positive. The following statement addresses the issue very well.

> We can do only what we think we can do. We can be only what we think
> we can be. We can have only what we think we can have. What we do,
> what we are, what we have, all depend upon what we think.
>
> Robert Collier

Many years ago, a young engineer attended a short course—part of his lifelong learning program—at a well-known university. The course was taught by an individual who was not particularly brilliant, but he was very self-assured. He simply exuded confidence and was intent on getting his message across. During the course, one of the members of the class asked a very good, but difficult, question. Everyone could tell immediately that the instructor did not know the answer. However, he took a very positive approach and began a straightforward engineering analysis of the question. He kept talking about every aspect of it as he assimilated all the issues that surrounded it. As he continued, he kept eliminating the extraneous points until he finally arrived at a logical conclusion, which everyone had to agree was unquestionably the correct answer. The class was fascinated with his techniques. Undoubtedly, if he had approached the question with a negative attitude, he would have thrown in the towel immediately, or his negative attitude would have persuaded him that he was unable to solve the problem. Without a doubt, your subconscious works with you to solve problems as long as you provide the impetus for it by maintaining both a positive attitude and the will to succeed.

Successful individuals inherently exhibit a positive mental attitude. This personality trait, which gives them a bold and optimistic approach to life, colors everything they do. Because of their attitude, they generate many ideas and, perhaps more importantly, they put these ideas into action.

Life is a game of inches. Quite often just a little more makes an enormous difference; therefore, the individual with the positive approach and can-do attitude will win. As an analogy to illustrate the importance of a small difference, engineering students learn quickly that water heated to 211° F is just hot water, but water heated to 212° F is steam, which can be used to drive turbines, which in turn drive generators, which produce electricity.

The benefits you obtain from a positive attitude are enormous. Your attitude not only influences your own actions, but it also influences the actions of those people with whom you come in contact. Your attitude is like a beacon to them; it not only influences them to follow you in some endeavor or support your position on some issue, but it is a solvent that dissolves some of the obstacles that hinder working relationships. It promotes health and happiness, and, as such, is a formidable shield against stress and illness. In addition, it gives you both energy and confidence to accomplish whatever task is at hand. There is a powerful, yet unknown, force that supports this attitude, and people who won't take "no" for an answer usually end up getting a "yes."

Many people have experienced occasions in which they took a chance, started a project, and assumed that eventually they would manage to circumvent all the obstacles in their path. In an almost miraculous manner, all the necessary ingredients fell into place, and they were able to accomplish whatever they set out to do. On the other hand, they have then found that, when they have been too cautious (because of fear or doubt), someone bolder than them has achieved what they had sought to do.

In the following poem, the unknown author makes the case for a positive mental attitude very well.

> If you think you are beaten—you are.
> If you think you dare not—you don't.
> If you'd like to win, but you think you can't,
> it's almost a cinch that you won't.
>
> If you think you'll lose—you've lost.
> For out in the world you'll find,
> success begins with a fellow's will.
> It's all in the state of mind.
>
> If you think you're outclassed—you are.
> You've got to think high to rise.
> You've got to be sure of yourself,
> before you can ever win a prize.
>
> Life's battle doesn't always go
> to the swifter or faster man,
> but sooner or later, the man who wins
> is the man who thinks he can.

Your success while you are in school, and after school as well, is critically dependent on the faith and trust you have in yourself, because, as stated in the poem above, if you don't think you can—you can't.

CAREER COUNSELING
AND PLACEMENT CENTER

Some colleges refer to this organization as simply the Placement Office; others use different, yet similar, titles. In any case, it is the office on each college campus that is responsible for helping you launch your career.

Purpose

While it is your responsibility to secure the employment you desire by convincing an employer that "you need me because . . . and here are some darned good reasons why!" your Career Counseling and Placement Center (CCPC) can be of enormous aid in the job-seeking process. The CCPC is organized and structured to serve as the critical catalytic agent to a successful interaction between you and the employers who have needs for individuals with your interests and talents. As the name implies, the CCPC exists to provide counseling services for all aspects of career planning, as well as an environment in which you and a prospective employer can meet to discuss the possibilities of future employment.

Services Provided

A well-organized and efficient CCPC can offer you a wide variety of services that significantly enhance your search for employment. The services will typically include, but are not necessarily limited to, the following:

- Individual counseling on such topics as strategies for a successful job search, résumé preparation, interviewing skills, and plant visit preparation.
- Workshops on writing letters or résumés, interview preparation, networking, and preparing for the start of a new job.
- On-campus interviews with representatives from business, industry, and government, including both public and private as well as large and small organizations.
- Job announcement bulletins that list job opportunities nationwide.
- Open résumé books, which contain the résumés of all students seeking employment, are available to potential employers. Even employers who do not visit the campus can request the résumés of students who meet their job requirements.

- Career fairs that permit students to learn about a wide variety of employment opportunities by interacting with many employers in a single setting over a short interval of time.
- Other services, such as company literature, books, and videos on topics that enhance the placement process, and computerized data banks on available jobs.

Registration

In order to participate in on-campus interviews, you have to register with the CCPC. This registration should take place when you are within about a year of graduation. However, it is very important that you realize that by far the best interviewing period is the October-to-April time frame. If you try to interview for a job outside this time interval, you will probably encounter great difficulty, since few companies visit the campus outside this time.

CCPCs, as a part of their placement services orientation, will insist that you attend a placement registration seminar in which all the rules and regulations governing their operation and your requirements are explained. For example, it is strictly verboten to miss a scheduled interview. Failure to adhere to the CCPC's rules is grounds for them to withdraw your interviewing privileges.

Qualities Employers Seek

From an employer's perspective, there are a number of qualities that they would like for every employee to possess. The better you understand the qualifications they are looking for, the better prepared you can be to exploit your inherent abilities and the good qualities you have developed over the years.

Intelligence is a primary prerequisite; however, if you are graduating in engineering, you have demonstrated this quality by virtue of being in a position to interview. Initiative is also an important trait. If your get-up-and-go has got-up-and-gone, you are not as desirable an employee as you could be. Dedication, demonstrated by a willingness to work hard, is not only an excellent quality, but it also serves as a good example for others. An ability to communicate clearly, both orally and in writing, is a real asset. Although you should be self-sufficient and capable of independent work, you should also be a good team player and operate in a courteous and diplomatic manner. In today's world, you have to be flexible. Things change quickly, and you must adapt. Finally, the more results-oriented you are, with an ability to separate the wheat from the chaff, the more potential you will demonstrate.

Self-Preparation

Your preparation for seeking employment should include a critical self-assessment of your likes and dislikes, strengths and weaknesses, and any other personal traits and characteristics that would affect your employment choices. You should also have your career goals firmly in mind. Clearly, some of the qualities listed here cannot be determined from an interview or an application form. They must be gleaned from performance on the job. However, the better you understand these factors, the more successful you will be.

Your résumé should be prepared to reflect your career goals. The placement office will have good examples of the format to use in preparing this document, and you should take advantage of their expertise in this area.

As you begin the interview process, do your homework on the potential employers. Know as much about them as possible before the interview. For example, what is their business, where are they located, what is their financial posture, what are the advancement possibilities, and the like. Furthermore, be prepared to explain your interest in their products and services, how you feel you can enhance their success, and why you are uniquely qualified to contribute to their future.

During an interview, be yourself, be prepared, be honest, be confident, be enthusiastic, and be professional in every aspect. However, you should also be realistic—if the job described is not suitable for your talents and interests, you should say so. If you are invited on a plant trip (a company-paid trip to examine their facilities and meet their personnel), all the items outlined here apply in that situation also. However, don't take the trip if you are not interested in the job!

LIFELONG LEARNING

Although the same can be said for individuals in any number of professions, it is critically important that engineering graduates continue to learn after graduation. The faculty at the university you attend will do its very best to prepare you for an engineering career. However, engineering is not a stagnant profession. Through research and development, constant advances in technology are being made every day. As an example, 35 years ago students studied vacuum tubes. However, since that time, electronics has progressed to transistors and then the placement of them in small scale integration (SSI) packages, medium scale integration (MSI) packages, large scale integration (LSI) packages, and very large scale integration (VLSI) packages. The socket area used for one of the vacuum tubes would now be occupied by hundreds of millions of transistors. This one example indicates why it is generally believed that the half-life of an engineering degree is approximately four to five years.

In addition to the rapid technological changes that have taken place, there are a number of changes that have affected engineering, such as the trends toward flattening organizations, growth in smaller companies, and the global nature of the business environment. Because the products that engineers specify, design, and develop will be sold in a worldwide marketplace, the engineers best prepared are those who learn elements of other engineering fields as well as the attendant areas of business, law, marketing, and government policy.

The tremendous changes that regularly occur in technology not only represent an exciting opportunity for engineers, but also a challenge to maintain their technical skills. However, technical vitality requires planning. Therefore, it is important for an individual to begin early to accept a personal responsibility for his or her careers, and an important aspect of this is lifelong learning. It is also interesting to note that, during the course of engineering careers, individuals may experience one or more job changes, and therefore it is impossible for an undergraduate program to provide all the knowledge necessary for them to remain competitive throughout their entire careers. Thus, in order to be educated for life, and not just the very first job, students should exit an engineering program knowing *how* to learn, because their enjoyment of their work as well as their survival as competent engineers will depend on it.

Even before you graduate from the university, plan to adopt a program of lifelong learning that will support your career development.

There are several ways to keep yourself up to date. First of all, if your job is challenging, you will need to keep abreast of the advances in the technology in which you work. You can also attend professional meetings, which are held locally, nationally, or internationally, where the recent developments in your profession are presented. There are normally a number of technical publications that deal with specific areas of technology. There are also universities and companies throughout the country that offer short courses in essentially every aspect of engineering. These courses may last from a couple of days to a couple of weeks. The courses typically concentrate on some specific aspect of technology and are designed to bring you up to speed quickly.

Whatever mechanism you choose to use, be sure to maintain your competence in your profession. Keeping yourself competent in your work not only leads to greater enjoyment of it, but your competence is perhaps your best security. If you do not keep yourself up to date, you will be run over or left behind. Furthermore, you will find within a relatively short time that the new engineering graduates will be talking about technology that is completely foreign to you.

PROFESSIONAL REGISTRATION

To prepare to become a licensed professional engineer or a licensed land surveyor in any state, one must pass two exams given by the National Council of Examiners for Engineering and Surveying (NCEES). The first exam is an eight-hour exam called the Fundamentals of Engineering (FE). An individual who passes this exam and graduates from an ABET-accredited curriculum is classified as an engineer-in-training (EIT) and therefore becomes a candidate for registration as a professional engineer or land surveyor.

The FE exam tests your understanding of the following subjects: chemistry, dynamics, electric circuits, engineering economics, fluid mechanics, mathematics, material science, mechanics of materials, statics, and thermodynamics. Because these subjects are typically covered in the first three years of an engineering program, the end of the junior year or first part of the senior year appears to be an optimum time to take this exam. Colleges of engineering have all the necessary information on the exam and, in fact, a university is normally one of the sites where the exam is periodically given.

An engineer-in-training who has at least four years of experience in engineering work, the quality of which indicates competence to practice engineering as a professional, may then apply to take the professional engineering (PE) exam. This exam is focused on determining competence in a specific discipline (e.g., chemical, civil, electrical, etc.). Upon successful completion of this exam, one is awarded a professional engineering license in the state in question. While there are other paths to becoming registered as a professional engineer, the one described here is typical for individuals who have entered an engineering school. The procedure for becoming a licensed land surveyor is similar, although not exactly the same.

Professional engineering in any state falls under the jurisdiction of The State Board of Registration of Professional Engineers and Land Surveyors. The headquarters for this organization is typically located in the state capital, and that office is a repository for all the rules and regulations that govern their activities.

To become a registered professional engineer, licensed to practice the profession, is a very important and noble goal. The State Board of Registration regulates the practice in order to ensure that engineering is performed in a highly professional and ethical manner; safeguards life, health, and property; and promotes the public welfare. The *Engineers' Creed*, which is the guiding light for professional engineers, states the following:

ENGINEERS' CREED

As a professional engineer, I dedicate my professional knowledge and skill to the advancement and betterment of human welfare.

I pledge

To give the utmost of performance;
To participate in none but honest enterprise;

To live and work according to the laws of man and the highest standards of professional conduct;

To place service before profit, the honor and standing of the profession before personal advantage, and the public welfare above all other considerations.

In humility and with need for Divine
Guidance, I make this pledge.

Although a large segment of the engineering population is licensed, many engineers are not. For example, the chief engineer in a company must be registered, because that individual must approve all engineering done by the firm. Because of the nature of their work, computer majors represent a large group that typically does not become registered. Nevertheless, regardless of the type of engineering work you become involved in, you should not needlessly close off options to future careers. Licensing protects the general public, and this protection issue is often thought to be the litmus test for the need to become registered.

In the preceding paragraphs, the terms *license* and *registration* have been used interchangeably. However, strictly speaking, the two words do not mean exactly the same thing. "Registration" is typically defined as a recording or listing in some official register, while "license" indicates a legal right to act that has been awarded by some authority that is competent to grant the permission. Thus, within the present context, the word "license" is the more accurate word. Hence, the National Society of Professional Engineers (NSPE) and the National Council of Examiners for Engineering and Surveying (NCEES) are taking steps to ensure that the word "license" is used in place of the word "register" in pertinent literature. (See the article entitled PEs Take Challenge: Just Say "Licensed." By Molly Galvin, *Engineering Times*, NSPE, Volume 17, Number 11, November 1995.)

One of the latest developments in professional engineering registration is the new trend to require continuing education units (CEUs) to renew a professional engineering license. Once again, these new requirements are necessary in order to ensure that those practicing the profession remain competent and up to date.

GRADUATE SCHOOL

Degrees

An alternative to going straight to work is graduate school. However, graduate school is not for everyone. First of all, you must qualify to enter. This is typically translated into having at least a B average in your course work while an undergraduate. Second, you must be mentally prepared. This means you must have a desire to continue as a student and have a willingness to forgo a good salary, with the expectation that you will eventually make up in the long term what you lose in the short term. Furthermore, it may lead you to a career that you could not pursue without this continued study.

There are actually two degrees at both the master's and doctoral levels. There is a Master of Science (M.S.) and a Master of Engineering (M.Eng.), and there is a Doctor of Philosophy (Ph.D.) and Doctor of Science (Sc.D.). Typically, the Master of Engineering degree is a nonthesis degree and is obtained by taking course work beyond the bachelor's level. The Master of Science degree requires a thesis, and the doctoral degrees require a dissertation.

The master's degree will normally take from one to two years to complete. The doctoral degree will typically take two to three years beyond the master's degree. A master's program tends to deepen your understanding of the field and help you put the pieces together. It increases your confidence by giving you a much better grasp of the technology. The doctoral degree is a research-oriented degree and therefore one that leads to a career in research and development or a university professorship. While it is extremely important for both of these professions, it is essentially a "union card" for the latter.

Paths to Graduate Education

There are many ways to pursue graduate school. For example, you can go from undergraduate school directly to graduate school. You can go to work for a company that agrees to send you back to school full-time. Not many companies will do this, and if they do, they are bound to require you to work for them for some period of time. You can also go to graduate school part-time while working for a company. This part-time approach is facilitated by the fact that many universities, including the National Technological University (NTU), offer a master's degree through a videotape-based program. This seems to be the best of both worlds—a good salary and graduate school, too. However, a word of caution is in order here. The company will look upon your graduate school as a luxury. If they need you to work overtime, or if they need to send you out of town for some period of time, graduate school will be the item that suffers; and in general, they will tell you that, while they hate to interfere with your schoolwork, work comes first. In other words, sometimes this approach just doesn't work. Finally, you can go to graduate school in a foreign country. Because of the global nature of many engineering businesses, graduate school in a foreign country would appear to be excellent preparation for a career in this industry. With this approach, you encounter two immediate problems—the language and financial aid. However, it is an excellent opportunity to visit a foreign country and broaden your collegiate exposure. Furthermore, if foreign graduate study is of interest to you, you could prepare for it by taking a foreign language in high school and continue it through your undergraduate program.

If you decide that graduate school is a viable option for you, then you must make a number of decisions—some of which may be made for you for a variety of reasons. For example, what area do you wish to pursue—microelectronics, composite materials, wastewater treatment, or what? If your grades are excellent, you may have your pick of the schools that have good programs in your area of interest; otherwise, you may have a limited number to choose from. However, keep in mind that decades ago if you wanted to go to a graduate school that had excellent professors in some area, you had to go to one of the best schools in the country. Today, there are many schools that have some outstanding professors. While it is a good idea to attend graduate school at a different university than where you earned a bachelor's degree, your own alma mater is also a good possibility. After all, you already have a track record with them.

Within the group of schools that accept you, there are many other factors to consider such as the financial aid package—fellowship, research assistant, teaching assistant—and geographical location (can you live where it's real cold or very hot, do you have transportation, or do you have to rely on public transportation, and, if so, does it exist—in many small university towns it is typically very limited).

Information Sources

There are many sources from which you can glean a lot about the schools you wish to consider. For example, Peterson's Guides are an excellent source. Their mailing address is

Peterson's Guides
P.O. Box 2123
Princeton, NJ 08543-2123
(609) 243-9111
http://www.petersons.com

In addition, the American Society for Engineering Education (ASEE) publishes a list of graduate schools with some detailed information, and the *U.S. News and World Report* has a special issue on colleges and universities. Your university library will have all three of these publications. However, an excellent source of data is your own department. Your professors will know professors at other schools who are working in your area of interest. The department chairman or head will also have access to a wide range of information that will be helpful to you in making a decision. You may even have the opportunity to remain where you are.

The Application Process

Once you have begun to focus on a limited number of schools, you should contact them to get the required application forms. If you call them or use E-mail or fax, they will normally put the packet in the mail immediately. You must also apply to take the Graduate Record Exam (GRE) and, in some cases, the Test of English as a Foreign Language (TOEFL). The TOEFL is required only for students whose native language is not English. Both exams are administered by the Educational Testing Service (ETS). The graduate school of your university will be able to provide you with application forms. You can also write or call ETS directly at the following address.

ETS
P.O. Box 6000
Princeton, NJ 08541-6000
(609) 921-9000
http://www.collegeboard.org

The general GRE is given every couple of months throughout the school year. Some programs may require a field-specific test, and these are typically given only twice per year.

You should also arrange for your school to send official transcripts of your grades to those graduate schools where you are applying.

When you have completed the application forms, arranged to take the entrance tests required, and requested official transcripts, you should begin to line up your references. Your professors may or may not remember you—after all, they see a very large number of students in a year. Therefore, when you select your references, take them a copy of your résumé and remind them of the specific course(s) you took from them. In addition, take along a self-addressed, stamped envelope, and tell them the time frame in which they should send their letter. Ask them to send the letter to all the schools you are interested in. They will be annoyed if you ask them to send a letter to one school, then go back weeks later for a letter to another school, and so on. Get your ducks in a row, and do it all the first time. As the deadline for the letter approaches, gently remind them of the deadline, and thank them for all their help.

When you send in your application for graduate school, ask for financial aid in the form of fellowships, research assistantships, teaching assistantships, tuition waivers, or whatever the school might offer.

Once you have done everything you can, then, after a reasonable length of time, call the school to check the status of your application to be sure that they have received everything that you have arranged to be sent to them. Finally, find out if any of your present professors know professors in the department where you wish to enter graduate school. If they do, ask them if they would be willing to call one of their colleagues and put in a good word for you.

If you are interested in attending graduate school at the university where you are enrolled as an undergraduate, you may be able to shortcut the process, because they can easily examine your transcript, and letters of reference are not needed because they already know you.

JOB SECURITY

Although job security is only an issue after you have completed a program in engineering, it has ramifications for the mind-set that must be developed when pursuing an engineering curriculum.

Today's industry, in order to remain competitive, must be very quick to adapt to what would appear to be the *one* constant in their business: *change.* Failure to adapt quickly can, in some cases, have very serious consequences. As a result, the engineering workforce must also be fleet-of-foot in the technology arena; that is, they must be willing and able to adopt a posture of continuous learning so that their skills are up to date and their tools are sharp.

Because knowledge and capabilities are key resources, they are always in vogue. Furthermore, they are easily transportable in what has become a global economy. Within this fast-shifting environment, it is your competence that is in demand, and it is your competence that is your best security. Thus, as you pursue a program in engineering, think in terms of learning everything that you possibly can, not in terms of obtaining an engineering degree. If you are tenacious in the pursuit of the former, you will automatically achieve the latter.

CHANGING PROFESSIONS

If you have your heart set on becoming an engineer, you might find it interesting to know that some people enter an engineering program knowing full well that they will never practice the profession. They use an engineering degree as a stepping-stone into another profession.

In many professions—for example, law and medicine—you have to have an undergraduate degree to be admitted into the professional program. The undergraduate degree does not have to be in pre-law or premed. In fact, in these professions, as well as many others, the professional schools are delighted to have students with an engineering degree apply. Students with engineering degrees are typically intelligent, hardworking, and disciplined, and they possess a much better understanding of the physical world than students who have gone through the preprofessional programs. For example, consider the mechanical engineering student who wants to be an orthopedic surgeon. The student's background in fluid and solid mechanics would be an obvious advantage. Or, the student who wants to be an attorney and specialize in environmental law. A background in civil engineering with its attendant courses in environmental issues, such as wastewater treatment, would be most advantageous.

It is not unusual for students to pursue another career after graduating in engineering. Either they decide that engineering is not for them, or they wish to use the technical base in some other endeavor. Engineering students have gone into agriculture, business, education, law, medicine, social work, as well as many other professions. In general, if they have the other skills necessary, engineers are academically prepared to pursue almost any profession.

11

Profiles of Careers

THE SURVEY GROUP

In an attempt to provide you with some ideas of what engineers do when they graduate with a bachelor's degree in engineering, a number of graduates throughout the country who have been out of school less than 18 months were contacted and asked to briefly describe their initial career path. In the material that follows, you will see the tremendous diversity of the types of activities in which many of the young graduates are involved. To aid you in this presentation, the group has been subdivided by engineering major, and a number of the disciplines outlined in Chapter 1 have been included.

CHEMICAL ENGINEERING GRADUATES

When Jerry graduated, he went to work for a paper mill in the southeastern part of the country. He is a process engineer. He works in a technical group comprised of six engineers. Their job is to get as much production as possible out of the mill. Projects are assigned individually, and he is responsible for his tasks. The company has only one mill, and therefore there is no travel associated with his job. He typically works 40 hours per week. Jerry likes his job and plans to stick with it. For recreation, he is an avid tennis player.

Monica is a process engineer with a large oil-refining company. She works with about a dozen other engineers, but is the only process engineer in the group. Her job is to ensure that the wastewater treatment facility meets the permit requirements for wastewater discharge. Most of her traveling is involved in taking continuing education courses. She typically works about 45 hours per week. She enjoys her work and plans to move around within the company to increase her experience. In her spare time, she likes to go to the beach.

Anthony graduated with a very high GPA. He came from a poor family, but had excellent grades and was practically put through school on scholarships. He worked at summer jobs to supplement his income while going to school. His last summer job was with a major drug manufacturer, and he found that he really liked doing creative work. So he opted to go to graduate school and is currently attending one of the most prestigious universities in the country.

Harry is a sales engineer for a large chemical company that manufactures a large array of chemicals. He spends many hours on the road. Harry may spend as much as a month at a customer's location, running tests and trials in order to sell his product. He normally takes some technical support with him in the form of two other engineers. He typically works about 60 hours per week. He hopes to either move up in the management of his company or start his own business. In his spare time, he likes to play bridge.

Ginger graduated from the university with a high GPA. Although she planned to enter engineering practice, she had always had in the back of her mind that perhaps she would like to be a medical doctor; therefore, she decided that if she was ever going to try it, the best time would probably be right away. She entered a medical program in the southwest and is doing very well. Although she misses engineering, she believes that she will be able, one day, to combine her knowledge of both areas in a viable career. She claims that she does not have a lot of spare time.

CIVIL ENGINEERING GRADUATES

Bridget graduated from the university with a very high GPA. She is an assistant project director with a state department of transportation (DOT).

She is out on the job site (e.g., a bridge construction project), in a hard hat and boots from 7 a.m. to 4 p.m. It is her job to supervise the state workers on-site and the other construction companies that work for the DOT. She is personally responsible for the construction to be done correctly. She travels a lot locally and typically covers the entire county in which she lives. She normally works 40 hours per week and is paid for any overtime. In her spare time, she has been involved in judging science fair projects in the local school system, and she likes to swim. Her plans for the future involve going to graduate school.

George's title is that of civil engineer. His company is a small engineering firm with about 150 people (25 percent of whom are engineers) that does engineering work in the electrical, architectural, and mechanical areas. George's specialty is the design of wastewater distribution systems. He normally works in a team on big projects and if the project is primarily one of a wastewater nature, he will act as the project leader. If the project is small, then he will work alone. He travels on average about two days per month and works about 50 hours per week. His future plans involve going back to school to get an MBA. He is an avid golfer and consistently shoots in the high 70s.

Amy is project engineer with a civil and environmental engineering consulting firm. She is involved in what she calls "hard-core design." She designs wastewater treatment facilities and is also concerned with the environmental impact statements for landfills and storm water systems. The company has about 150 employees, and on small jobs she typically works with one other engineer. Her job involves both office and field work. A lot of her office work involves computer modeling for water transmission systems. The amount of traveling she does depends totally on the project and, in some cases, can involve quite a bit. She typically works about 50 hours per week. She loves her work and plans to stay with it. She does, however, want to get her Professional Engineer's license and move up to project manager. In her spare time, she exercises a lot and participates in church activities. At the time of the contact, she was planning a wedding and was totally immersed in that.

As Charlie went through his undergraduate program, he decided, based on some of his course work, that he wanted to work in the environmental engineering area. The school that he attended did not have a program in environmental engineering at the undergraduate level, so he decided to pursue a master's program in that area. He is currently about halfway through the program at a major university, working as a graduate teaching assistant. He is enjoying his studies and his teaching. He is paid a modest stipend to teach, but it is sufficient for his needs. He is looking forward to graduating in about a year and entering the workforce in the environmental area.

Clifford works for a small company (about 100 employees) that sells mixtures for concrete. He is in technical sales and is a lone ranger in that he works by himself. He does have some service people who back him up on occasions when their help is needed. He travels anywhere from 2,000 to 3,000 miles per month within the state where he lives. In his normal routine, he works about 55 hours per week. Clifford's goal is to have a business of his own. In his spare time, he likes to hunt, fish, and play tennis.

ELECTRICAL ENGINEERING GRADUATES

Lisa used the cooperative education program to get through school. Although she did her co-op work with a government contractor, when she graduated from school, she went to work for a major electric utility. She liked her co-op work and believes that it gave her a good idea of what to expect when she began working full-time in the real world. She designs the control systems, which employ programmable logic controllers, for the generation of steam in fossil-fuel power plants. She normally works by herself, but has a support group that works with her on certain projects. She does the design in the office and then the installation is, of course, done in the plants. She will typically travel four to five days per month. She plans to get as much experience as possible in the design and instrumentation area and is currently not thinking beyond that point. In her spare time, she enjoys camping and traveling.

Joe was a co-op student while in school and maintained a high GPA. His co-op assignment was with a local telephone company. His father is a physician and could have easily paid for Joe's education, but did not do so because he believed that Joe would get more out of school if he worked his way through. During his senior year, Joe took an elective course in image processing and became fascinated with the subject. He decided that he would like to go to graduate school in that area. He was interested in applying image processing to medicine, with techniques such as magnetic resonance imaging (MRI). Joe is now a graduate research assistant at a major university, working with a professor whose research is in the image processing area. When he has a break in his schoolwork in the winter, he loves to snow ski and can do so with ease. In the summer, he likes to fish.

When Ian graduated from the university, he went to work with a friend of his father's. He has known the man most of his life. He is in the technical sales area and works as a manufacturer's representative. Ian covers an area of approximately a 200-mile radius, including two relatively large cities. He typically works with electronic measurement equipment and goes to customers in his area to demonstrate it. He does a lot of traveling and typically works about 55 hours per week. He enjoys hunting and fishing in his spare time and hopes to learn enough in his work to start his own business.

Sam's education in the area of electrical engineering was unusual from the very beginning. His father had a small law firm in a relatively unpopulated Midwestern area. His father wanted Sam to enter the law firm and eventually take it over when his father retired. Sam was interested in doing just that. So why did Sam go into electrical engineering if he was planning on a career in law? The answer is simply that his father had told him that he would need an undergraduate degree in some area before he could enter law school, and he felt that a degree in electrical engineering would teach him to think critically and analyze problems in a systematic manner. So Sam is now in law school at a different university and planning on joining his father upon graduation.

Fran's title is that of engineer, and she works for an engineering firm that derives the majority of its business from government contracts. She is involved in the design of infrared weapon systems and in her work uses a wide variety of software programs. She typically works by herself; however, because of her co-op experience gained while going through school, she also heads up a group that is involved in training other people in the company. Her traveling is confined to attending continuing education courses that keep her abreast of new technology in her field, and this traveling averages no more than one or two days per month. There are times when she will work 70 hours per week (that includes 13 hours per day on the weekend) while she is involved in supervising other contractors who work through her company for the government. She has just had a baby and is interested in obtaining an aerobic instructor's certificate. If she can get this certificate, she can teach aerobics after working hours, and the place where she would teach has a day care center where she can leave her child. In the future, she would like to pursue a master's degree in electrical engineering.

MECHANICAL ENGINEERING GRADUATES

Juan works for a mechanical contracting firm, and his title is that of engineer. The firm is small and does about $15 million worth of business per year. Juan estimates that his time is divided as follows: 70 percent of the time he is involved in estimating the cost of a job (i.e., bid specification), 10 percent of his time is spent in purchasing and making sure that equipment arrives at the job site on time, 10 percent of his time is spent in design, and the remaining 10 percent is spent in the CAD area with shop drawings. He normally works by himself, and on large jobs he is the key person responsible for all the air distribution and duct work. He enjoys what he does and feels that he is exposed to a wide range of design firms from all over the country. He does very little traveling and typically works about 50 hours per week. He plans to get his Professional Engineer's license and hopes to one day be a partner in the firm. For recreation, he likes to play tennis.

Teresa is called a general engineer and works with one of the U.S. government laboratories that does research on wood for paper products. She is involved in testing and analyzing the mechanical properties of paper, specifically its coefficient of friction. Her work impacts paper manufacturers and the printing industry. The laboratory has about 300 people, and she works in a group of about 10 people that moves from product to product to solve various problems. She typically works under one of the more senior engineers and normally works about a 40-hour week. She does very little traveling in her job, but what traveling she does is confined to continuing education courses and conferences. Her job is a fixed-time contract and will end after a three-year period. She wants to move to a position where she can do work in heating, ventilating, and air conditioning. At present, she is taking a correspondence course in preparation for taking the Fundamentals of Engineering exam; later, she would like to obtain a Professional Engineer's license. In her spare time, she is involved in all types of recreational activities.

Adrienne is a boiler performance engineer with a utility company. She travels throughout the company's service area testing boilers in power plants to make sure that they meet the clean air requirements. She is one member of a team of four engineers that attempts to balance efficiency and emission control. The engineering department of the company has hundreds of engineers, and there are groups such as the one that she is in whose jobs include testing turbines. Depending on the type of job in which she is involved, she may be the lead engineer or she may act in a support capacity. She does a lot of traveling. She is away from home, on average, about three months out of the year. It takes about three to four weeks to do a test, and therefore she is gone for this period. However, she is home on the weekends. Adrienne typically works about 40 hours per week unless they are very busy, in which case she may work 50 to 60 hours per week. She enjoys what she is doing and plans to stick with it, learning all that she can about the business. She is an outdoors-type person and enjoys playing basketball. At the moment, she is planning a wedding, and that activity is taking up most of her time.

Ryan works with a small engineering firm that has about 30 employees. The company is owned by a much larger company that does environmental engineering work. Ryan is involved in abatement and remediation work. He oversees the removal of such things as asbestos, PCBs (polychlorinated biphenyls), heavy metals, and aromatic hydrocarbons. Although he works in a team and reports to a project supervisor, he is responsible for his part of the project. In his work, he follows the job and relocates wherever the job is; that is, when he finishes one job, he moves to another. He normally works about 50 to 55 hours per week. At present, he has no plans for the future. He likes what he is doing and feels that he is too young to worry about planning too far ahead. He likes the outdoors and spends most of his spare time hunting and fishing.

Brad worked his way through school as a diesel mechanic. He worked for a very large trucking firm that had hundreds of tractors and trailers. When he received his degree, he was made the director of maintenance for the firm. As such, he is responsible for keeping all the rigs running. He also works very hard to control maintenance costs, keep all the electronics on the trucks in working order, and is the individual who is responsible for matching the most efficient tractor to trailer when purchasing new equipment. In addition, he is the person they call if there is an accident and diesel fuel is spilled. He supervises about 70 technicians who work in a number of depots over a multistate area. He typically makes about one trip per month in his supervisory position and attends professional meetings that deal with the trucking industry. He normally works about 45 to 50 hours per week, unless there is a diesel fuel spill somewhere in the country. He has passed the FE exam and is currently working toward his Professional Engineer's license. He plays golf with customers, and in his spare time he is a member of a softball league.

BIOMEDICAL ENGINEERING GRADUATES

When Paul was in high school, he thought that he might like to be a medical doctor. So he asked some of the physicians in his hometown if he could go with them while they visited with some of their patients. They were happy to accommodate his wishes, and, as a result, Paul decided to pursue a career as an orthopedic surgeon. During a visit to the university he planned to attend, he found that they had a program in biomedical engineering. He entered the program and took his elective courses in biomechanics. He is now in medical school pursuing a program in orthopedic biomechanics. He loves his studies, but has little or no time for anything else.

Curtis had always been interested in medicine, but did not want to practice as a medical doctor. He learned that the university had a program in biomedical engineering, and so he entered that program with the hope that he would be able to work for a drug company. When he graduated, he accepted a job with a large drug manufacturer and is currently involved in the development of delivery systems for genetically engineered drugs and screening tests that determine the applicability of this type of drug therapy. He enjoys his work and normally works a little more than 40 hours per week. His passions are skiing and backpacking, and he does these things every chance he gets.

COMPUTER ENGINEERING GRADUATES

Because Eileen's father was a computer systems programmer, she had been exposed to computers for a long time prior to entering the university. She decided that she wanted to pursue a similar career, and therefore she entered a program in computer engineering. While in school, she was much more interested in software than hardware and, as a result, when she graduated, she took a job with a major defense contractor as a software engineer. She does programming in a real-time environment for a weapon system under development for the military. She normally works about 40 hours per week unless there is a design review, in which case she may work overtime in preparation for her presentation. She enjoys playing golf in her spare time.

Keith was a little older than most students when he entered a program in computer engineering. He joined the Navy upon graduation from high school and worked as a radio operator. He then went to work for a major electronics manufacturer as a commercial electronics technician. When he graduated from the university, he went to work for a government contractor. He is involved in the development of a Windows-based automated testing system that controls complex communications test equipment and shipboard radio equipment. He has been able to replace some outdated equipment with some of the modern technology that is capable of efficiently handling the task with cheaper, more reliable equipment. His work schedule typically requires about 40 to 50 hours per week. He enjoys water sports and owns a small sailboat that he uses every chance he gets.

ENVIRONMENTAL ENGINEERING GRADUATES

Tim's family was always concerned about environmental issues, and he guesses that it was this influence that guided him toward a career in environmental engineering. When he graduated from the university, he went to work for a consulting firm that specializes in municipal wastewater treatment. He was paired with an experienced engineer involved in the design and construction of a new wastewater treatment system. In this capacity, he works with both the laboratory technicians to determine the characteristics of the wastewater and the state to obtain a discharge permit. He is also part of a team that must decide what types of processes must be linked together to achieve the level of treatment desired at the cheapest cost. His work schedule is approximately 45 hours per week, sometimes longer when he is in the field. Tim is really into backpacking and does it every chance he gets.

When Sheila entered the university, she was not really sure what curriculum she would follow. However, one of her girlfriends was going into environmental engineering, and so Sheila thought she would look at that curriculum closely as a possible career path. After talking to some of the professors in that area, she decided to enter that program. When she graduated, she took a job with a state agency that is involved in permitting activities for industrial and municipal wastewater treatment systems. Sheila calibrates a waste load allocation model of a receiving stream in order to set discharge limits for a proposed wastewater treatment system. She rarely works overtime, loves her job, and spends her free time exercising and working as an aerobics instructor.

INDUSTRIAL ENGINEERING GRADUATES

Melanie's father was an industrial engineer, and he had convinced her that this would be a good career path for her. After graduation, she went to work for a large semiconductor company on the west coast. She works in a team that is designing the layout of a manufacturing process. They are concerned with efficiency and the control of production and materials. Melanie is personally responsible for the ergonomic aspects of the jobs and must ensure that each job is safe and that any risk of injury is removed. She enjoys her job and typically works about 40 to 50 hours per week. In her spare time, she likes to play tennis and has joined a health club where she can play and meet other people with similar interests.

When Allen graduated from the university, he joined a company that manufactures snack foods. He is on a one-year training assignment as a production supervisor. In this capacity, he is responsible for all aspects of the production line, including dealing with personnel problems, scheduling the materials, and maintenance. He will soon complete this assignment and looks forward to going into some engineering function with the company. Although Allen works a 40-hour shift, his work normally extends far beyond this time due to problems that invariably arise. When he has time to do so, Allen is an avid bicycle-racing enthusiast.

12

Two Final Thoughts

The previous sections of this book have tried to cover some of the issues that will directly or indirectly determine your ultimate success, and in particular your success as a student, in an engineering curriculum. There are, however, a couple of general comments that will be helpful to you as you approach a program in engineering. The first comment deals with the definition of success. As the famous British writer H. G. Wells has said:

> The only true measure of success is the ratio between what we might have been on the one hand, and the thing we have made of ourselves on the other.

Note carefully that this statement makes no reference to comparing yourself with your fellow students. Furthermore, it does not imply that you will earn an A in a specific course or that you will graduate with honors. However, success, as defined here, requires discipline. Success costs, and you have to pay the price; unfortunately, it is not cheap. You are free to choose from among the many opportunities with which university life presents you, and through a disciplined approach, you are able to match, as best you can, your talents and interests to the opportunities at hand.

The final comment is this. As you approach an exciting and reward-ing career in engineering, through an educational program that contains numerous courses and laboratories that are defined by terms you perhaps do not even understand at this time, you should be guided by the words written by M. Louise Haskins and spoken by King George VI of England in his 1939 Christmas message to the British Empire:

> I said to the man who stood at the gate of the year, "Give me a light, that I may tread safely into the unknown," and he replied, "Go out into the dark-ness and put your hand into the hand of God. That shall be to you better than light and safer than a known way."

Appendix

The Accreditation Board for Engineering and Technology, Inc.
  (ABET)
111 Market Place, Suite 1050
Baltimore, MD 21202-4012
(410) 347-7700
http://www.abet.ba.md.us

PARTICIPATING BODIES

American Academy of Environmental Engineers (AAEE)
132 Holiday Court, #206
Annapolis, MD 21401-7005
(202) 296-2237
http://www.asee.org/

American Congress on Surveying and Mapping (ACSM)
5410 Grosvenor Lane

Bethesda, MD 20814-2144
(301) 493-0200
http://www.techexpo.com/tech_soc/acsm.html

American Institute of Aeronautics and Astronautics (AIAA)
370 L'Enfant Promenade SW
Washington, DC 20024-2518
(800) 639-2422
http://www.lmsc.lockheed.com/aiaa/sf/home.html#hq

American Institute of Chemical Engineers (AIChE)
345 East 47th Street
New York, NY 10017-2395
(212) 705-7338
http://www.che.ufl.edu/aiche/

American Nuclear Society (ANS)
555 N. Kensington Avenue
LaGrange Park, IL 60526-5592
(708) 352-6611
http://www.ans.neep.wisc.edu/

American Society of Agricultural Engineers (ASAE)
2950 Niles Road
St. Joseph, MI 49085-8607
(616) 429-0300
http://www.asae.org

American Society of Civil Engineers (ASCE)
345 East 47th Street
New York, NY 10017-2330
(800) 548-2723
http://www.asce.org

American Society for Engineering Education (ASEE)
11 Dupont Circle
Washington, DC 20036
(202) 331-3500
http://www.asee.org/asee

American Society of Heating, Refrigeration, and Air-Conditioning
   Engineers, Inc. (ASHRAE)
1791 Tullie Circle, NE
Atlanta, GA 30329-2305
(404) 636-8400
http://www.ashrae.org/

The American Society of Mechanical Engineers (ASME)
345 East 47th Street
New York, NY 10017-2330
(800) 843-2763
http://www.asme.org

Institute of Industrial Engineers (IIE)
25 Technology Park
Norcross, GA 30092-2901
(770) 449-0461
http://www.iienet.org

The Institute of Electrical and Electronics
   Engineers, Inc. (IEEE)
345 East 47th Street
New York, NY 10017-2330
(800) 678-4333
http://www.ieee.org/

The Minerals, Metals and Materials Society (TMS)
420 Commonwealth Drive
Warrendale, PA 15086-7511
(412) 776-9000
http://www.tms.org

National Council of Examiners for Engineering
   and Surveying (NCEES)
P.O. Box 1686
Clemson, SC 29633-1686
(803) 654-6824
http://www.ncees.org/ncees

National Institute of Ceramic Engineers (NICE)
757 Brooksedge Plaza Drive
Westerville, OH 43081-2821
(614) 890-4700
http://www.acers.org/

National Society of Professional Engineers (NSPE)
1420 King Street
Alexandria, VA 22314-2750
(703) 684-2800
http://www.nspe.org

Society of Automotive Engineers (SAE)
400 Commonwealth Drive
Warrendale, PA 15086-7511
(412) 776-4841
http://www.sae.org

Society of Manufacturing Engineers (SME)
One SME Drive, Box 930
Dearborn, MI 48128-2408
(313) 271-1500
http://www.sme.org

Society for Mining, Metallurgy, and Exploration, Inc. (SME-AIME)
8307 Shaffer Parkway
P. O. Box 625002
Littleton, CO 80162-5002
(303) 973-9550
http://www.smenet.org

Society of Naval Architects and Marine Engineers (SNAME)
601 Pavonia Avenue
Jersey City, NJ 07306-2907
(201) 798-4800
http://www.sname.org/

Society of Petroleum Engineers (SPE)
P.O. Box 833836
Richardson, TX 75083-3836
(214) 952-9393
http://www.spe.org/

AFFILIATE BODIES

American Consulting Engineers Council (ACEC)
1015 Fifteenth Street, NW, STE 802
Washington, DC 20005-2670
(202) 347-7474
http://www.acec.org/

American Institute of Mining, Metallurgical and Petroleum Engineers (AIME)
345 East 47th Street
New York, NY 10017-2330
(212) 705-7695
http://www.smenet.org/aime.stml

American Societies for Nondestructive Testing, Inc. (ASNT)
1711 Arlingate Lane
P.O. Box 28518
Columbus, OH 43228-0518
(614) 274-6003
http://www.asnt.org/ndt

American Society of Safety Engineers (ASSE)
1800 East Oakton Street
Des Plaines, IL 60018-2187
(708) 699-2929

Instrument Society of America (ISA)
67 Alexander Drive
P.O. Box 12277
Research Triangle Park, NC 27709-2277
(919) 549-8411
http://www.isa.org

Society of Plastics Engineers (SPE)
14 Fairfield Drive
Brookfield, CT 06804-0403
(203) 775-0471
http://www.bbsnet.com/spe

References

1. Claire LeBuffe, "Showing Declines—Enrollments '93," *ASEE Prism Magazine*, September 1994, pp 35–37.
2. Peter F. Drucker, "The Age of Social Transformation," *Atlantic Monthly*, November 1994, pp 53–80.
3. Mark Alpert, "The Care and Feeding of Engineers,"*Fortune Magazine*, September 21, 1992, pp. 86–95.
4. Minutes of the ASEE Engineering Deans' Council Business Meeting, Toledo, Ohio, June 23, 1992.
5. Alfred J. Engel, Ed. *From Microchips to Potato Chips: Chemical Engineers Make a Difference*, New York: AIChE.
6. *Our Past, The Present, Your Future . . . In Civil Engineering*, New York, ASCE.
7. *Your Career in the Electrical, Electronics, and Computer Engineering Fields,* Washington, DC: IEEE United States Activities Board.
8. *Mechanical Engineering, A Career for the Future*, New York: ASME.
9. *Engineering and You*, New York: NSPE, 1991.

10. "Degrees '92: A Steady Output," *ASEE Prism Magazine*, February 1993, pp. 24–26.

11. *Alabama Law Regulating Practice of Engineering and Land Surveying*, Montgomery, AL: State of Alabama Board of Registration for Professional Engineers and Land Surveyors, 1991.

12. *U.S. News and World Report*, September 28, 1992, pp. 96–127.

13. *1990 Annual Report*, New York: ABET, 1990.

14. *Auburn University Placement Manual*, Evanston, IL: CRS Recruitment Publications, 1992.

15. *Manpower Comments*, vol. 29, no. 9, December 1992.

16. Dale Carnegie. *How to Stop Worrying and Start Living*, New York: Simon and Schuster, 1948.

17. Dale Carnegie. *How to Win Friends and Influence People*, New York: Simon and Schuster, 1964.

18. Napoleon Hill and W. Clement Stone. *Success Through a Positive Mental Attitude*, New York: Pocket Books, 1977.

19. Donald M. Dible, Ed. *Build a Better You—Starting Now*, Fairfield, CA: Showcase Publishing Company, 1979.

20. Norman Vincent Peale. *Why Some Positive Thinkers Get Powerful Results*, Nashville, TN: Foundation for Christian Living in cooperation with Thomas Nelson Inc. Publishers, 1982.

21. James C. Gardner, "Three Lessons for Living," *Reader's Digest*, March,1987, pp 185–187.

22. *Engineering Education and Practice in the United States*, Washington, DC: National Academy Press, 1994.

23. Private Communication, Chairman K. Ward, Department of Freshman Engineering, Purdue University, West Lafayette, IN.

24. Molly Galvin, "PEs Take Challenge: Just Say 'Licensed,'" *Engineering Times*, NSPE, vol. 17, no. 11, November 1995.

Go in peace. The journey on
which you go is under the eye of
The Lord.

Judges 18:6

Index

About the Author

J. David Irwin was born in Minneapolis, MN, on August 9, 1939. He received the B.E.E. degree from Auburn University, Auburn, AL, in 1961, and the M.S. and Ph.D. degrees from the University of Tennessee, Knoxville, in 1962 and 1967, respectively.

In 1967, he joined Bell Telephone Laboratories, Inc., Holmdel, NJ, as a member of the technical staff and was made a supervisor in 1968. He joined Auburn University in 1969 as an assistant professor of electrical engineering. He was made an associate professor in 1972, associate professor and head of the department in 1973, and professor and head in 1976. In 1993, he was named Earle C. Williams Eminent Scholar and Head.

Dr. Irwin has served the IEEE Computer Society as a member of the Education Committee and as Education Editor of the magazine *Computer*. He has served as chairman of the Southeastern Association of Electrical Engineering Department Heads and the National Association of Electrical Enginnering Department Heads and is past president of both the IEEE Industrial Electronics Society and the IEEE Education Society. He is a life member of the IEEE Industrial Electronics Society AdCom and has served as a member of the Oceanic Engineering Society AdCom. He served for two years as Editor of the *IEEE Transactions on Industrial Electronics*. He has served on the Executive Committee of the Southeastern Center for Electrical Engineering Education, Inc. and was president of the organization in 1983/1984. He has served as an IEEE Adhoc Visitor for ABET Accreditation teams. He has served as a member of the IEEE Educational Activities Board and was the Accreditation Coordinator for IEEE in 1989. He has served as a member of numerous IEEE committees, including the Lamme Medal Award Committee, Fellow Committee, and the Nominations and Appointments Committee. He has served as a member of the Board of Directors of the IEEE Press. He has also served as a member of the Secretary of the Army's Advisory Panel for ROTC Affairs and Nominations Chairman for the National Electrical Engineering Department Heads Association. He has been and continues to be involved in the management of several international conferences sponsored by the IEEE Industrial Electronics Society.

He is author and co-author of more than 50 publications, papers, and presentations, including *Basic Engineering Circuit Analysis,* first through fourth editions published by Macmillan and the fifth edition published by Prentice-Hall. He is also co-author of four other textbooks, *An Introduction to Computer Logic, Industrial Noise and Vibration Control, Introduction to Electrical Engineering,* and *Digital Logic Circuit Analysis and Design,* which are published by Prentice-Hall. He is the Editor-in-Chief of *The Industrial Electronics Handbook,* in production with CRC Press.

He was made a Fellow of the IEEE in 1982 and received an IEEE Centennial Medal in 1984. He was awarded the Bliss Medal by the Society of American Military Engineers in 1985. He received the IEEE Industrial Electronics Society's Anthony J. Hornfeck Outstanding Service Award in 1986, was named IEEE Region III (Southeast) Outstanding Engineering Educator in 1989, and in 1991 he received a Meritorious Service Citation from the IEEE Educational Activities Board, the 1991 Eugene Mittelmann Achievement Award from the IEEE Industrial Electronics Society, and the 1991 Achievement Award from the IEEE Education Society. In 1992, he was named a Distinguished Auburn Engineer. In 1993, he received the IEEE Education Society's McGraw-Hill/Jacob Millman Award. He is a member of the American Society for Engineering Education. In addition, he is a member of Sigma Xi, Phi Kappa Phi, Tau Beta Pi, Eta Kappa Nu, Pi Mu Epsilon, and Omicron Delta Kappa.

About the IEEE Press

As the book publishing arm of the IEEE, Press is devoted to quality and growth. Today's Press is publishing and copublishing outstanding texts and references in many areas of electrical and electronics engineering, including:

- Professional handbooks for on-the-job applications
- Practical guides to career issues such as management, communications, and professional development
- Cutting-edge monographs on fast-growing areas in the field
- Applied engineering texts of current technologies and practices
- Graduate textbooks for advanced students and professionals

Our top quality Press staff, hired from major international publishing houses, has years of successful experience in editorial development, marketing, production, and finance.

To see the full selection of IEEE Press titles, visit the IEEE Virtual Bookstore via our Home Page at Http://www.ieee.org. IEEE members receive generous discounts on IEEE books and other product purchases. If you would like to become an IEEE Member, call 1-800-678-IEEE (Toll-Free USA and Canada) or 1-908-981-0060. If you have an idea for proposing a book of your own, contact the IEEE Press editorial offices at 908-562-3967 or e-mail ieeepress@ieee.org for an author's kit.u